5

Topics in Medicinal Chemistry

Lead-Seeking Approaches

Volume Editor: Matthew M. Hayward

With contributions by

J. A. Bikker · J. W. Ellingboe · K. D. Freeman-Cook ·
A. M. Gilbert · R. K. Harrison · D. W. Kung · C. A. Lipinski ·
J. E. Morin · L. S. Narasimhan · J. W. Paslay · M. Pellecchia

Springer

Editor
Dr. Matthew M. Hayward
Pfizer Global Research & Development
Dept. Neuroscience Chemistry
Eastern Point Rd.
Groton CT 06340
MS 8220/4339
USA
matt.m.hayward@pfizer.com

ISSN 1862-2461 e-ISSN 1862-247X
ISBN 978-3-642-01074-3 e-ISBN 978-3-642-01075-0
DOI 10.1007/978-3-642-01075-0
Springer Heidelberg Dordrecht London New York

Library of Congress Control Number: 2009935068

Cover design: KünkelLopka GmbH, Heidelberg, Germany
Typesetting and Production: SPi Publisher Services

Printed on acid-free paper

Springer is part of Springer Science+Business Media (www.springer.com)

Volume Editors

Dr. Matthew M. Hayward

Pfizer Global Research & Development
Dept. Neuroscience Chemistry
Eastern Point Rd.
Groton CT 06340
MS 8220/4339
USA
matt.m.hayward@pfizer.com

Editorial Board

Topics in Medicinal Chemistry
Also Availabe Electronically

Topics in Medicinal Chemistry is included in Springer's ebook package *Chemistry and Materials Science.*

If you do not have access, you can still view the table of contents of each volume and the abstract of each article by going to the SpringerLink homepage, clicking on "Browse by Online Libraries", then "Chemical Sciences," and finally by selecting *Topics in Medicinal Chemistry.*

You will find information about the

- Editorial Board
- Aims and Scope
- Instructions for Authors
- Sample Contribution

at springer.com using the search function by typing in *Topics in Medicinal Chemistry.*

Color figures are published in full color in the electronic version on SpringerLink.

Aims and Scope

Drug research requires interdisciplinary team-work at the interface between chemistry, biology and medicine. Therefore, the new topic-related series *Topics in Medicinal Chemistry* will cover all relevant aspects of drug research, e.g. pathobiochemistry of diseases, identification and validation of (emerging) drug targets, structural biology, drugability of targets, drug design approaches, chemogenomics, synthetic chemistry including combinatorial methods, bioorganic chemistry, natural compounds, high-throughput screening, pharmacological in vitro and in vivo investigations, drug-receptor interactions on the molecular level, structure-activity relationships, drug absorption, distribution, metabolism, elimination, toxicology and pharmacogenomics.

In general, special volumes are edited by well known guest editors.

In references *Topics in Medicinal Chemistry* is abbreviated *Top Med Chem* and is cited as a journal.

Preface to the Series

Medicinal chemistryisbothscienceandart. Thescienceof medicinal chemistry offers mankind one of its best hopes for improving the quality of life. The art of medicinal chemistry continues to challenge its practitioners with the need for both intuition and experience to discover new drugs. Hence sharing the experience of drug discovery is uniquely beneficial to the field of medicinal chemistry.

The series Topics in Medicinal Chemistry is designed to help both novice and experienced medicinal chemists share insights from the drug discovery process. For the novice, the introductory chapter to each volume provides background and valuable perspective on a field of medicinal chemistry not available elsewhere. Succeeding chapters then provide examples of successful drug discovery efforts that describe the most up-to-date work from this field.

The editors have chosen topics from both important therapeutic areas and from work that advances the discipline of medicinal chemistry. For example, cancer, metabolic syndrome and Alzheimer's disease are fields in which academia and industry are heavily invested to discover new drugs because of their considerable unmet medical need. The editors have therefore prioritized covering new developments in medicinal chemistry in these fields. In addition, important advances in the discipline, such as fragment-based drug design and other aspects of new lead-seeking approaches, are also planned for early volumes in this series. Each volume thus offers a unique opportunity to capture the most up-to-date perspective in an area of medicinal chemistry.

Dr. Peter R. Bernstein
Prof. Dr. Armin Buschauer
Prof. Dr. Gunda J. Georg
Dr. John Lowe
Dr. Hans Ulrich Stilz

Preface to Volume 5

The journey of a drug from conception to launch is a long road, fraught with many detours and dead ends. Compound attrition is caused by a myriad of sources, including suboptimal safety, potency and/or PK/PD. In the last decade, the industry has started focusing more on the identification of quality lead matter that possesses a multidimensional package of attractive properties rather than advancing hits solely on potency. Hits with lead quality attributes and devoid of potential toxic pharmacophores advance faster through the discovery pipeline and stand a better overall chance of success.

In this volume we examine how hits are typically identified and validated in the pharmaceutical field and how this chemical matter is advanced to lead stage. The introductory chapter by Chris Lipinski is a commentary on the role of medicinal chemists in this process, based on his long and successful career in the industry. The following three chapters speak of hit identification using traditional high throughput screening techniques (Paslay, Morin and Harrison), virtual screening (Narasimhan and Bikker) and NMR screening techniques (Pellecchia). The final two chapters describe the hit triage (Freeman-Cook and Kung) and the follow-up processes (Ellingboe and Gilbert) for advancing hits into leads.

The objective of this volume is to provide an overview of the hit-to-lead process; it is not intended to be all inclusive. Furthermore, much of the methodology described here is influenced by the culture of the authors' respective institutions. In general, however, lead seeking approaches across industry (and academia) are more similar than they are different and are, for the most part, captured in this volume.

I am indebted to the authors for their contributions to this volume. From each of their respective chapters I have learned something new, unexpected and pertinent to my own career in drug discovery. I am also indebted to the many readers here at Pfizer who contributed a commentary on the chapters and helped to influence the final version presented here.

July 2009 Matthew M. Hayward

Contents

Top Med Chem (2009) 5: 1–24
DOI: 10.1007/7355_2009_4

Published online: 30 June 2009

Overview of Hit to Lead: The Medicinal Chemist's Role from HTS Retest to Lead Optimization Hand Off

Christopher A. Lipinski

Abstract A medicinal chemist combines organic synthesis expertise and the ability to optimize chemistry structure–activity relationships (SAR) based on relevant biomedical information so as to achieve project goals. The ability to optimize chemistry SAR consists of both the easier to explain logical stepwise structural modification that is often described by quantitative structure–activity relationships (QSAR) and the more difficult to explain exercise of high-order pattern recognition. Optimizing SAR is full of traps for the unwary. What are the pros and cons of various types of screens? Should one believe the screening data? How does one optimize against multiple sometimes conflicting properties? What types of compounds are worth screening? How does one judge the quality of a screening hit? Very importantly, drug discovery is a team exercise in which the medicinal chemist plays a key facilitating role. Given good interpersonal skills, the medicinal chemist's training is broad enough to enable cooperative interactions across the whole discovery team. Chemistry pattern recognition is the unique skill that the medicinal chemist contributes to drug discovery. The ability to relate chemistry structure to biological activity and to change chemistry structure so as to change a variety of biomedical parameters in a desired direction leads to the successful "drug hunter."

Keywords HTS, Hit to lead, Medicinal chemistry, Pattern recognition, Screening

Contents

C.A. Lipinski(✉)
Scientific Advisor, Melior Discovery, 10 Connshire Drive, Waterford, CT 06385-4122, USA
e-mail: clipinski@meliordiscovery.com

Abbreviations and Symbols

<	Less than
>	Greater than
ADMET	Absorbance distribution metabolism excretion toxicity
Apo structure	Structure of a protein without a bound ligand
ATP	Adenosine triphosphate
Cherry pick	Individually selecting a screening sample from a larger set
CLND	Chemi luminescent nitrogen detector
CNS	Central nervous system
DMSO	Dimethylsulfoxide
DNA	Deoxy ribonucleic acid
DOS	Diversity-oriented synthesis
GPCR	G-protein coupled receptor
H-bond	Hydrogen bond
HERG	Human ether-a-go–go related gene
HIV	Human immunodifficiency virus
HPLC	High performance liquid chromatography
HTS	High throughput screening
IC50	Inhibitory concentration (at) 50% inhibition

IP	Intellectual property
kg	Kilogram
L	Liter
Log *P*	Logarithm (base 10) of *P* (Partition coefficient)
MDR1	Multidrug resistant (Gene) 1
mg	Milligram
mL	Milliliter
mM	Millimolar
MS	Mass spectrum
MWT	Molecular weight
NDA	New drug application
NIH	National institutes of health
nM	Nanomolar
NMR	Nuclear magnetic resonance
NSAI	Non steroidal anti inflammatory
P	Partition coefficient (of drug) between normal octanol and water
PCR	Polymerase chain reaction
PDE	Phospho diesterase
pK/pD	Pharmacokinetics/pharmacodynamics
QSAR	Quantitative structure – activity relationship
RO5	Rule of 5
SAR	Structure – activity relationship
sdf	Structure data file
SiRNA	Short intervening ribonucleic acid
SPR	Surface plasmon resonance
μM	Micromolar
z′	HTS quality criterion based on positive and negative compound controls

1 Introduction: What is a Medicinal Chemist?

The author has spent many years as a medicinal chemist in the pharmaceutical industry and is taking this opportunity to share a broad range of advice and opinions. First of all; what is a medicinal chemist? A medicinal chemist is not the same as a synthetic organic chemist. Consistent with the years of experience cited for expert witnesses in litigation an expert medicinal chemist might be expected to have at least 10–15 years of relevant biomedical pattern recognition superimposed on a solid synthetic organic synthesis background. The pattern recognition is the linking of biomedical information to chemistry structure. Chemical structures associated with emotionally significant events (compound activity success or failure) are stored in the medicinal chemist's amygdala and are instantly available for retrieval. This is an example of a very high order of pattern recognition found in humans (and other mammals) that is evolutionarily selected for because of its survival value. Most highly skilled professions exhibit some sort of very high order of pattern recognition as exemplified in the book "Blink" by Malcolm Gladwell [1]. This pattern recognition is mostly a blessing but occasionally a problem. The blessing is that this skill is at the core of medicinal chemistry competency. The problem is that this skill is very difficult for non-chemistry professionals to understand. In particular, biologists may not understand how a skilled medicinal chemist can make a "snap" and accurate judgment about a compounds quality simply by viewing the compounds chemical structure.

Two distinct patterns are evident in the US-based hiring of medicinal chemists. The major pharmaceutical companies hire the best synthetic chemists they can find and then teach them the medicinal chemistry part on the job. A second medicinal chemistry hiring pattern is for a smaller organization to hire a person with an academic medicinal chemistry background. For the very best people after about 15 years the overall skills profile from both hiring patterns may be very similar.

1.1 The Medicinal Chemist's Role

The medicinal chemist's role is to optimize all the aspects of the properties of a drug so that the compound succeeds in the clinic. Most often discussions focus on the technical aspects of "hit to lead." However, it is important to consider the following. What is the most troublesome; technical issues or people issues? Always people issues are more troublesome and more difficult to solve than technical issues. Although this chapter is mostly about technical issues the reader would be advised to remember the trouble priority order. The medicinal chemist to be effective functions as part of a team. The team aspect always exists but intensifies along the progression "hit to lead" towards "lead optimization." Cooperative interactions with genomics, biology, drug metabolism and pharmaceutical sciences are essential. The medicinal chemist needs to be aware that the single most important factor in success is the chemistry starting point since the structures of many drugs (but not the properties) may not differ greatly from that of the original hit [2]. People related factors come in to the choice of a hit. Is the biological assay valid? Was there some assay problem related to poor compound properties that colored the choice of hit? Is there some complexity in the biology that early on should have been a chemistry warning flag about pursuing a hit? Dealing with questions like these requires both some relevant biology background as well as interpersonal skills in dealing with the biology discipline. The opportunities for miscommunication operate in both directions. The medicinal chemist may have problems with understanding the biology. Equally important, the facility with which the nonchemistry disciplines understand chemistry varies greatly and will generally be poorest in the genomics/biology disciplines and better in pharmaceutical sciences and drug metabolism. In communicating with drug discovery sister disciplines it is well to remember that it is not good enough to give the correct message. The message has to be formatted so that it is understand by the audience. This may well mean that the chemistry issues that are so clear to the medicinal chemist will need to be very clearly explained to genomics/biology colleagues without using chemistry jargon.

Specifically what can a medicinal chemist do to enhance cross discipline communication? In dealing with biologists always ask to see the chemistry structure. Do not wait. Eventually you will ask for the structure and so asking earlier rather than later will save you a lot of aggravation. In dealing with biologists try to convey the point that when you make a compound it will have a definite rather than an indeterminate structure. The biologist is quite happy to talk about error limits but you the medicinal chemist really need to get some idea as to which structure is more active because this is so important for your next synthesis. In dealing with pharmaceutical scientists try to get the information in graphical format. Most medicinal chemists get limited information from the equations so loved by pharmaceutical scientists. Almost any pharmaceutical science equation can be converted into an easily understandable graphic.

2 Drug Discovery, Druggability and Developability

Let us assume the goal in medicinal chemistry is to work on a drug discovery target. Do we know the structural and physicochemical characteristics of successful drugs? The answer is a resounding YES. As potential drugs progress from preclinical to phase 1, phase 2a and 2b, phase 3, NDA and final market approval there is a very consistent change in properties. Parameters such as MWT and Log P decline throughout the process such that the mean approved drug MWT is less than 350 and less than 15% of launched drugs have log $P > 5$. Multiple literature reports come to this conclusion. Remarkably, the profile of clinically approved drugs has remained very stable for 40 years despite the increased MWT and Log P of compounds entering phase 1 [3–5]. This very time stable trend strongly suggests that in some way relatively simple properties track with development success (i.e., developability). The developability stands in contrast with druggability, the ability of a ligand to bind to target with properties consistent with oral activity and adequate pharmacokinetics and pharmacodynamics. The physicochemical profiles of ligands as a function of target are now well known. For example, against aminergic GPCR's, kinases, PDE's, most ion channels the vast majority of ligands will be RO5 compliant. However against targets like aspartyl proteases and GPCR's having peptides or lipids as natural substrates the vast majority of ligands will be non-RO5 compliant. With a great deal of effort, skill and access to the appropriate resources well resourced companies have pushed against the fringes of RO5 space and successfully entered compounds into the clinic with adequate oral absorption and pk/pd (e.g., HIV protease inhibitors). This then leads to a conundrum. The industry has been initially successful in entering compounds into the clinic against the challenging (and very interesting targets). With enough resources and effort ligands can be found that have the credentials to enter the clinic, i.e., they are druggable. But so far there is no evidence that the historical pattern of approval success in relatively small and nonexcessively lipophilic compounds has changed. In other words the newly developed ligands are outside of the historical norms of approved drugs. While they may be druggable they are only poorly developable.

2.1 High Throughput Screening

The predominant method in the current era of finding hits against a target is HTS. As defined by Wikipedia this is "Using robotics, data processing and control software, liquid handling devices, and sensitive detectors, HTS allows a researcher to quickly conduct millions of biochemical, genetic or pharmacological tests." The "millions" may be a bit of an exaggeration since the most common format today is to test compounds in plates containing either 384 or 1,536 wells. The vast majority of current HTS consist of the search for a selective ligand against a single mechanistically defined target. Recently this paradigm has come under increasing criticism and there is an increasing diversity in how one can discover leads [6]. The reader should be aware that there are valid alternatives (with associated pros and cons) to HTS and that many drugs were discovered without resorting to HTS.

Currently there is a slow gradual growth in the pursuit of poly-pharmacology [7, 8]. This can be done by mixing of multiple compounds each with a mechanism that is synergistic to the desired biology outcome or by seeking a single compound having multiple synergistic mechanisms of action. This latter approach has existed for decades (whether known or not) in the search for CNS drugs. In this therapeutic area it is the rule that clinically approved compounds typically exhibit a dozen activities at receptors/transporters in the ten nanomolar affinity range. The skilled medicinal chemist's pattern recognition has certain inevitable consequences when using HTS to discover leads. Most importantly, a skilled medicinal chemist can never design a truly diverse screening library. This is a good thing because the literature overwhelmingly indicates that biologically active compounds are not uniformly distributed in chemistry space [9–11]. A logical follow up to this observation is that screening a truly diverse library is the worst way to discover a drug. A skilled medicinal chemist often in collaboration with a computational chemist is essential in screening library design because the medicinal chemist's prejudices and biases will always be productively incorporated into the library design.

2.2 Phenotypic Screening

Screening for phenotype rather than mechanism was the norm in the 1970s, an era at least as productive as the current in terms of drug approvals [12]. Screening for mechanism predominated by the 1990s and still does today. Most drugs entering into the clinic in the 1970s arose from phenotypic (observational) in vivo screens, for example blood pressure lowering, lipid effects, glucose lowering, rodent behavioral assays. The common pattern was that there was no mechanistic bias behind the phenotypic end point. This resulted in a huge advantage in terms of target opportunity space. For example, as a minimum there are at least 100 possible mechanisms that might be responsible for a hypoglycemic phenotype. In the 1970s the mechanism was not known for most compounds entering the clinic. Knowing mechanism is still today not a regulatory requirement to enter clinical trials. However, not knowing mechanism does considerably complicate clinical development, for example especially on backup strategy. If something untoward happens in late stage discovery or in development the first question asked is whether the untoward effect is on target or off target. If the undesired effect is off target, then changing chemical structure usually solves the problem. Recently at least two major drug companies have reinstituted phenotypic drug screening as a complement to mechanism based drug screening, at least in part because of the increased target opportunity space of the phenotypic screen.

Medicinal chemists often raise objections to phenotypic screening. They do not believe that it is possible to optimize activity significantly against a multi parameter end point or a mechanistically undefined end point. This viewpoint comes from the experience with activity optimization against a single mechanistically defined end point. In this process it is common to improve in vitro activity by three orders of magnitude, e.g., moving IC50 from 1 μM to 1 nM. This medicinal chemistry argument is in some ways correct. It is indeed very rare to improve in vivo activity by three orders of magnitude. This happens because in vitro and in vivo activity do not scale in a linear fashion [13]. The reasonable expectation might be three orders of magnitude improvement in vitro and perhaps 10- to 20-fold improvement in vivo. The more modest in vivo activity improvement was in fact the norm in the 1970s for phenotypic screening in vivo in animal models. So the current medicinal chemists are not incorrect in their opinions but they are biased by their emphasis on in vitro activity. One has to ask, after all. What is the key optimization metric? Is it in vitro activity improvement or in vivo activity improvement?

Phenotypic screening is particularly common in academia where a natural product collection or diversity-oriented synthesis (DOS) compounds are screened. These types of compounds are narrow in ligand opportunity space in that they are unlikely to bind in multiple modes. Hence, they are a very good fit with the broad target opportunity space of a phenotypic screen. A phenotypically active compound is a wonderful starting point for the often difficult and time consuming mechanism deciphering and an excellent opportunity for publishable research by a graduate student or post doctoral fellow. In the current era, high content cell-based screens can be configured to be either mechanistically oriented or phenotypic in nature. With modern dyes and single cell level optical resolution it is possible to monitor cell phenotypic behavior, for example, movement of a cell surface feature or to see something translocating from cytoplasm to nucleus.

2.3 Fragment Screening

Screening low molecular weight fragments often in the range 175–300 is now a well validated method of discovering starting points for chemistry optimization. The rule of three for fragment screening suggests fragment libraries of compounds with MWT <300, log *P* <3, rotatable bonds and less than three H-bond acceptors [14]. It is common for a smaller drug structure to be embedded in another larger drug structure and experience teaches that small drugs are wonderful sources of fragments. The prejudice of a decade ago that low MWT compounds could not be optimized to a lead and to a clinical candidate status has now been proven incorrect. Fragments typically bind with affinities in the low hundreds of μM to low mM range. As a result the screening technique must be able to accommodate high concentration (and high aqueous solubility) to measure these types of weak affinities. Most commonly, various types of NMR assay have been used as well as X-ray methods with soaking in or crystallization of a ligand. The extra structural information from X-ray and often NMR greatly facilitates the optimization process. NMR and X-ray are specialist technologies that to some extent have limited the uptake of fragment screening. Additional technologies used either to complement NMR or X-ray or as replacements for these two techniques include: high dose biochemical assays, isothermal calorimetry, affinity-based MS, surface plasmon resonance (SPR) with immobilized protein target or SPR on immobilized tethered fragments. These technologies all have the commonality that the screening experiment provides less structural information than X-ray or NMR and hence the chemistry optimization becomes more difficult. Computational advances in the fragment arena include software to shred biologically active compounds into chemically sensible fragments; to recombine fragments into lead-like compounds and software to mutate fragment lists into new fragments having similar structural or pharmacophoric features. Fragment screening advantages are; better coverage of chemistry space with lower MWT ligands; steeper SAR for the optimization process and utility for probing the druggability of marginal targets. Disadvantages are; the need for specialist type assays such as NMR or X-ray capable of detecting weak fragment affinities in the 100 μM to low millimolar range and no real advantage for a clearly undruggable target.

The steeper SAR for fragment optimization derives from the role of enthalpic and entropic contributions to fragment binding. As a rough rule, the entropic energy penalty for binding of a fragment to a target is independent of size. Roughly half the potential binding energy of a small fragment might be lost due to unfavorable loss of translational and rotational entropy. As the fragment size is increased the entropic penalty stays roughly the same and the increased enthalpic binding energy component shows up as a steep activity optimization SAR. So it is not at all impossible to start with a 1 mM fragment binder and end up with optimization to a 1 nM binder in a relatively small number of steps. Fragment screening (to date) has shown no real advantage for the classic undruggable target (e.g., a protein–protein interaction). However, there are targets in the gray zone for which druggability is unclear. For these targets fragment screening offers an attractive exploratory technique.

2.4 Structure-Based Drug Design

Having an X-ray structure of a target is a tremendous advantage in chemistry optimization. As a rough rule, about 25% of the advantage resides in the apo structure (without a ligand). The full value resides when ligand is bound to target. Unfortunately, if the target is membrane bound as in GPCRs there is almost never structural information because of the universal difficulty in crystallizing membrane bound proteins. In predicting ligand binding to protein structures there are at least two important unsolved problems: those of protein flexibility and those of solvation. Currently it is very difficult to predict movements of large protein motifs from apo structure alone. Large protein motif movements in a target often uncover ligand binding sites, hence the value of target plus ligand information. Solvation changes in both ligand and target as a consequence of binding are very important to affinity but are currently rather poorly dealt with in computational chemistry. The computational prediction of docking (does the ligand bind to target?) and scoring (what is the relative affinity of ligand to target?) are currently evolving. Docking done properly is useful. Scoring is rather less precise with affinity rankings just a few years ago roughly paralleling MWT (higher MWT leads to higher affinity). Docking algorithms are steadily improving and their utility is enhanced by advances in liquid cherry picking robotics. Just a few years ago the ability to cherry pick samples from HTS plates or tube arrays was in the low 1,000s per day. Now in the best resourced companies it is possible to cherry pick samples in the tens of thousands per day. Moreover, advances in nanoliter dispensing offer the promise of high volume cherry picking moving into the affordable category for the smaller pharmaceutical companies and biotech companies. So rather than running docking and scoring to identify low hundreds of compounds for experimental screening, organizations of disparate size may soon be able to perform the same process on tens of thousands of compounds.

2.5 Homology Modeling

If the exact structure of a target of interest is not known it may be possible to construct computationally a model of the desired target from the known structure of a structurally related protein. As might be expected, the accuracy of a homology model declines as the sequence similarity of a known protein declines relative to the desired target. For some targets overall sequence similarity is much less relevant than the sequence similarity in the suspected ligand binding domain. Homology modeling is widely used and its accuracy is controversial. However, this author believes that the process can be useful even if the homology model is badly wrong. The interaction of a medicinal chemist with a homology model frequently leads to ways of thinking about structure that would not otherwise have happened and some of these insights do indeed lead to SAR breakthroughs. So my advice is not to be overly critical in advance of the accuracy of a homology model and I would advise being open to idea generation even if the homology is fairly low (e.g., 25–30%).

2.6 Chemoinformatics

In the HTS retest to lead optimization process it is essential that the medicinal chemist has access to chemoinformatics skills. Scoping out the intellectual property landscape is an essential part of synthesis planning. After all, it might not be a good idea to begin work on an area riddled with IP problems. All medicinal chemists need expertise in conducting exact structure, similarity, and substructure searches on specifically enumerated compounds in the patent literature and in peer reviewed journals. A Tanimoto similarity of 85% using a typical bit string structure description translates to something that a medicinal chemist would roughly characterize as structurally (but often not biologically) similar. The medicinal chemist should also be aware of recent US Supreme Court rulings that raise the bar to a compound being non-obvious. Older assumptions as to what constitutes non-obvious should be checked with internal experts. Competently run, patent Markush searches are not easy and are efforts best left to internal search experts. The trick in IP background searches is to narrow the search down to at most a few hundred references. Looking at abstracts alone, as well as specific compounds, will generally give a rough overview of the IP situation. Prediction of biological activity such as desired activity or toxicity from structure alone is in theory possible but practically speaking is still in its infancy. Some well resourced organizations have particular expertise in this area and consultation with internal experts is advised.

Some expertise in elementary virtual library generation is a plus. When deciding what to make next it is often advantageous to think about a group of compounds that might be made rather than a single compound. An easy method to generate a machine readable sdf file of virtual structures is a good starting point. This file can be run through whatever computational prediction programs are available and the output can then be mapped against whatever is known experimentally. The principle is that it is better to have more choices for synthesis than a single choice.

3 High-Order Pattern Recognition

There is a tendency to think of medicinal chemistry as primarily a logical exercise. A specific and trivial example would be the much maligned QSAR exploration of methyl, ethyl, butyl, futile. This author believes that equating medicinal chemistry with QSAR is incorrect. There is a definite place for what might for want of a better term be called high-order pattern recognition. A specific example is the time tested management strategy of dealing with a project stuck at an SAR impasse. In this situation experienced medicinal chemists from totally different projects are brought in. They look at the SAR with new, unbiased eyes. Maybe they see something in the chemistry that reminds them of something they worked on years ago. They see that the old chemistry can be applied to the new series and they make a compound that is an SAR breakthrough that the former project team would never have made and a compound that very possibly is not any kind of logical progression based on the preexisting chemistry. Amazingly enough this unexpected outcome process works well enough so that its use is repeated over and over again in knowledgeable drug discovery organizations.

3.1 HTS and the Defined Mechanism Screen

Screening compounds against a mechanistically defined biochemical or cell-based target is currently the predominant method of generating positives (actives) in drug discovery. The classic biochemical approach would be to try to discover a small molecule ligand that is a potent and selective antagonist of receptor "Z." This method has the advantage that by definition the mechanism of a true positive is known. The disadvantage is that the target opportunity space is low. For example, successful screening for a potent and selective antagonist "Z" gives just that: a potent and selective antagonist "Z." All the other mechanisms that might also give the desired biological effect are not explored.

3.2 HTS, Library Design and the Medicinal Chemist

Medicinal chemists ideally should be involved in screening library design. Often this might involve collaboration with a computational chemist. The goal is to take advantage of the skilled medicinal chemist's pattern recognition. The worst way to design a screening library is based on a "maximum chemical diversity" approach by a computational chemist who knows nothing about medicinal chemistry. In this author's opinion the best screening libraries incorporate prior biomedical knowledge. Since biological activity is so tightly clustered in chemistry space it makes sense to use this knowledge. For example, one approach might be to accumulate collections of biologically active compounds and then to shred them apart computationally in a chemically sensible manner. The fragments can then be computationally reassembled in new ways. This so called "recap" technology [15] is now available in commercially available software and sophisticated methods exist to depict fragments in "fuzzy" chemistry space so that new fragments can be generated from a preexisting fragment list. Experimentally, the biggest problem with this approach is that large regions of biologically active fragment space are not populated by commercially available compounds.

3.3 HTS True Positive Hit Rates

The percentage of true positives in an HTS varies widely depending on assay quality and on target type [16]. In a high quality HTS with z' greater than 0.5 (an accepted criterion of quality) the hit rate might vary between 0.1 and 2%. HTS hit rates are very subjective in that usually assay conditions are set so that the total number of alleged actives does not overwhelm the resources available for post HTS triage. Targets vary tremendously in druggability [17] from good; kinases, aminergic GPCRs to difficult but tractable; proteases, peptidergic GPCRs to near hopeless: phosphatases, protein–protein interactions. The target druggability is exactly mirrored by the ligand profile. Druggable targets yield RO5 compliant ligands which minimizes any difficulties in achieving oral absorption. Difficult targets yield ligand profiles mostly lying outside of RO5 space. Targets like protein–protein interactions are undruggable from an HTS perspective. The dozen or so examples of small molecule ligands for protein–protein interactions are exceptions to the undruggable generalization. There is something special about the target, for example a cavity or hot spot at the interaction surface that moves the target towards druggable (but likely with difficulty). Not surprisingly the more undruggable the target the higher the false positive hit rate. The number of compounds screened in an HTS can depend as much on budget as on science. The cost of an HTS is typically expressed in two ways; what it costs the screeners budget and what it costs the corporation. Expressed in screener budget terms the low end for a 384-well format HTS might be 5 cents per well. Expressed in terms of cost to the corporation, which includes a fair share of all costs, the price per well might be more like 50 cents. Clearly HTS costs mount quickly for multiple screens against million compound size libraries.

3.4 False Positives in HTS

The number of false positives always rises as a function of number of compounds screened. On a probabilistic basis there will always be a distribution of false positive actives at a certain number of standard deviations above the assay baseline of inactives. When people talk about false positives in this sense they are talking about data points that appear active because of the "noise" in the assay. In an HTS it is not uncommon for the ratio of false positives to true positives to be 5:1 or greater. The more undruggable the target, the greater the ratio and the more difficult is the post HTS triage – the process to figure what is a true positive from a false positive. The medicinal chemist has a major and critical role in this process. Generalizing, the role of computational chemistry is to reduce the number of apparent "positives" to a tractable number that the medicinal chemist can inspect. A single medicinal chemist manually examining the structures of several hundred compounds is doable. Most medicinal chemists would balk at manually looking at the structures of thousands of compounds.

Do medicinal chemists agree on compound quality? It depends on what is asked. If the desired input is binary: yes or no, good or bad then the agreement is not great. But this I think is the incorrect question. The question should be posed in two steps. Do you see a potential problem and how optimistic are you that you can fix the problem in chemistry? A specific example is illustrative. There would be pretty good agreement among medicinal chemists that a nitro aromatic moiety is a potential problem. With just a few exceptions nitro aromatics are mutagenic (more so in prokaryotes than in eukaryotes). The nitro group is reduced to very reactive species capable of undergoing covalent chemistry with nasty toxicity consequences. However, sometimes a nitro group is just a very electron-withdrawing group and the target biological activity is not related to any nitro group derived covalent chemistry. There are known bioisosteres for nitro groups that obviate the chemical reactivity and mutagenicity problems of the nitro group. So if the nitro group is functioning simply as electron-withdrawing it might be possible to replace it with a bioisostere such as an *N*-linked tetrazole. I think that most medicinal chemists would agree on the first step, identifying a potential problem but based on experience might differ on their degree of optimism on executing a chemistry fix. The bottom line here is that using a panel of experienced medicinal chemists to review HTS hits is better than relying on a single person.

3.5 Stochastic False Positives: Mostly in Biology

HTS false positives that are stochastic (random) are almost always related to the biology screen. One can construct a whole laundry list of causes: problems with reagents, problems with readers, problems with liquid handling, temperature control issues, mislabeling and translocation errors, HTS plate edge evaporation, etc. Because these errors are random they will not replicate. The general rates of concordance across HTS is known. For triplicate assays where the target, the compounds, and the assay format are allegedly the same the triplicate concordance is 50%, i.e., only 50% of the true positives will show up as active in all three of a triplicate assay. The concordance in a triplicate assay gets worse if the target and compounds are the same but the assay format changes. In theory, assay format should not matter but it does. If the assay readout changes in one of the triplicates the concordance declines to about 30–35%. What all this means is that reproducibility of replicates is an uncertain measure of even the true positives.

3.6 Nonstochastic False Positives: Mostly in Chemistry

HTS false positives that are nonstochastic (nonrandom) are mostly chemistry related. There is something about the chemical structure that leads to an HTS assay readout that is interpreted as an active. But this something is not related to activity at the biology target and the active is worthless as a chemistry SAR starting point. Another whole laundry list of causes can be identified: excessively lipophilic compounds that stick to plastics; detergents; cytotoxic compounds; compounds with optical properties that interfere with the assay; compounds interfering with a biology reporter readout; compounds acting at targets by way of colloidal aggregates [18] rather than as discrete compounds and redox compounds. The seriousness of the promiscuous nature of colloidal aggregates is significant with the incidence rising as screening concentration rises. At compound concentration of 30 μM the incidence could be as high as 20%. Compounds with redox activity are problems because it is common for a target protein to contain one or more disulfide linkages. Reducing such a linkage (or forming disulfide bonds from thiol functionality) can alter protein conformation and this might be interpreted as an assay positive [19].

The false positive compounds due to chemistry are insidious. Because the problem is related to the chemistry structure per se the worthless activity will reproduce. Such compounds are to be avoided in chemistry at all costs. If misfortune occurs and such a lead gets into chemistry optimization, the hallmark is a flat muddy SAR. Structural changes do not make much difference in activity. Compounds worth pursuing in chemistry show real SAR. Activity changes when chemistry structure changes. It is a good sign if all activity is lost when a methyl group is put in a wrong position. Biological activity should be sensitive to structural change. Virtually all drug discovery organizations have compiled rules and filters for removal of worthless compounds from HTS. The reason why compounds are worthless in HTS is not always known. Based empirically on experiment, organizations compile lists of worthless frequent hitter structural types. A specific example would be the condensation product of a benzaldehyde with an oxindole or anything structurally similar to the highly acidic heterocycles related to the NSAI Tenidap. Care must be exercised in identifying frequent hitters. There are the worthless structures, but there are also the worthwhile frequent hitters. These are structural motifs capable of chemistry SAR optimization recurring in compounds active at targets that have nothing in common in a biological sense. These are "privileged structures," and the benzodiazepine motif is such an example.

3.7 False Negatives

In general the rate of false negatives are by definition difficult to ascertain. There are two general approaches to get a handle on false negatives. The first approach is based on what is known about the aqueous solubility of screening compounds since truly active compounds out of solution are the most common cause of false negatives. One can infer that perhaps 15% of true positives will be missed in an HTS. This inference comes from an analysis of the concordance or lack of concordance between nominal concentrations in DMSO stocks and nominal concentrations on dilution into aqueous HTS assay buffers compared to actual measured concentration in DMSO stocks and actual concentration in aqueous HTS assay buffers. Generalizing, for about 85% of samples in HTS, the discrepancy between nominal and actual concentration is relatively small, i.e., the two values differ by a factor of 3. However for about 15% of samples the actual concentration will be markedly lower than the nominal concentration, an order of magnitude or more and frequently at the lower limit of quantitation. What has happened is that the compound has precipitated either from the DMSO stock solution or upon dilution into the aqueous HTS buffer. Compound insolubility is the single major cause of false negatives in HTS and it is a serious problem when IC50 rankings are based on nominal concentrations or when selectivity panel information is based on nominal concentration. Determining concentration experimentally without running standard curves is necessary with medium throughput or true HTS. Technically this is an analytical chemistry challenge. The best detector technology is that of CLND which has a reputation as a specialist detector technology. Success has been reported but so far only in expert hands. Adding experimental compound concentration to HTS assay data corrects for IC50 ranking errors and selectivity panel errors. It is this author's opinion that analytical chemistry monitoring of true compound concentration is the single most important technical advance that could improve HTS quality. This topic is the subject of current intense pharmaceutical industry exploration.

A second approach to handling false negatives relies on a computational analysis of actives in the primary HTS. Were there analogs or similar compounds to actives that appeared inactive in the original HTS? These inactives are retested perhaps in a more careful screen and some of the original inactives will now be found to be active. Most commonly these were truly active compounds that appeared inactive in the original screen because they were not in solution under the initial HTS assay conditions.

4 Screening: What is the Goal?

It is very important to understand the goal in an HTS screen because the screening goal has a very large impact on what types of compounds should be screened, on how many compounds should be screened, and how much the screening compounds might be expected to cost.

4.1 Chemical Biology/Chemical Genetics Screening

Chemical biology/chemical genetics is the use of a chemical compound as a tool or probe to learn something about a biology pathway [20]. The chemical compound is used in the same sense as a mouse knockout experiment or an SiRNA experiment, i.e., compound is used in a target validation sense. If a biological pathway is perturbed does something happen in biology? This is important because it is so frequent that perturbing a biology pathway causes nothing observable to happen (or nothing perceived to be useful to happen). Biological pathways are robust and designed by evolution not to be perturbed. In general, parallel pathways are least likely to be perturbed while pathways passing through a node are most likely to be perturbed. The traffic accident analogy is appropriate. Traffic is most likely to be disrupted if the accident occurs at a major intersection. Conversely, if an accident can be easily bypassed as by a parallel route, traffic will choose the bypass route.

In a chemical biology/chemical genetic sense the most important attribute is selectivity. Whatever the compound structure it must be selective enough so that something useful is learned in the experiment. The compound can contain chemistry flaws that would be unacceptable in a drug discovery setting. It just has to be selective and has to be compatible with assay conditions. This means that functionality with covalent chemistry possibilities may be acceptable. Specific examples might be the presence of warhead groups such as aldehydes and alpha halo ketones frequently used in exploration of protease pathways. As in any area, ambiguities exist. Is a boronic acid moiety acceptable only in a probe/tool compound or could it be permissible in a real drug (as in the proteasome inhibitor Velcade). Allowing the acceptability of screening compounds with covalent chemistry potential decreases the size of screening libraries. For a chemical biology/chemical genetics purpose a library of 50,000 compounds might be adequate for many target validation studies. A library of this size would be too small for real drug discovery against most targets. An exception would be directed libraries as one might use in the search for an ATP competitive kinase inhibitor where the likelihood of success would be expected to be high with a library of this size. In screening compounds price is in general an indicator of quality. Tool probe compounds which are frequently not desirable in real drug discovery tend to be lower priced. The pilot phase of the NIH Roadmap Molecular Libraries Screening Center Network effort searched for tools and probes. Many of the compounds in the screening libraries are poorly suited for drug discovery (but still quite adequate as tool compounds) hence the cost is low – $16 per 10 mg. By way of contrast, high drug discovery chemical quality translates into higher price. In 2005, Pfizer completed a major effort in screening library expansion, obtaining high drug discovery quality compounds. Designed overwhelmingly by medicinal chemists, these compounds were far more expensive reaching costs of $200–300 per 15 mg for some chemical series.

4.2 Drug Discovery Screening

Screens intended for real drug discovery require very high quality compounds. The universal rule is to filter out compounds with chemical flaws prior to screening rather than after screening [21]. Companies have learned by experience that prescreening is much the more efficient process. Presentations and publications from the major pharmaceutical companies on the culling of their legacy libraries all have the same theme. Typically one-third to half the legacy libraries are removed from screening because of a combination of functional groups to avoid, poor physicochemical properties and concerns about purity/identity. Drug discovery organizations have colorful acronyms to describe these drug discovery chemistry quality filters that illustrate how they view flawed compounds. The Vertex company's filters are termed REOS, Rapid Elimination Of Swill. Pfizer's comparable filters are called LINT, the worthless stuff collecting in the bottom of pockets or pant cuffs.

Library screening size has evolved over the last decade. With few exceptions the pattern has been to prepare smaller size libraries. In the early 1990s solid phase synthesis predominated over solution phase synthesis. The driver was the greater efficiency of solid phase synthesis if very large libraries were desired. The driver for the very large library was the misguided enthusiasm of senior pharmaceutical company executives (this authors opinion) for libraries in the thousands or tens of thousands in size. Screening experience gradually taught that a larger number of smaller libraries gave better screening results than a single very large library containing an equal number of compounds. As a result, solution phase synthesis with its advantages of speedier development time began to predominate. We now can understand this trend to smaller libraries as relating to ligand efficiency [22]. It turns out that the ligand efficiency of heavy atoms in a library core is usually higher than that of the heavy atoms in the appendages. Hence, other things being equal, 10 libraries of 100 compounds each (with 10 different cores) performs better than a single library (with 1 core) of 1,000 compounds. Very large screening libraries have not completely disappeared but now they are more commonly found associated with some specialist technology, e.g., assay signal amplification by PCR in DNA encoded libraries.

4.3 Compound Quality Filters Aka Functional Groups to Avoid

The drug discovery compound quality filters from about the year 2000 are in the public domain from Pfizer and Glaxo Wellcome (just before the merger with SKB). These filters have been recently summarized [23]. Organizations regularly update their filters but the changes mostly consist of addition of filters for infrequently occurring functionality. Hence the public domain filters serve as a useful guide. The reader is also referred to the excellent compilations of functional groups to avoid by Gilbert Rishton. For drug discovery quality, two types of filters are usually merged: namely those for unacceptable physicochemical properties (e.g., RO5 violation, excessive rotatable bonds) and filters for undesirable chemical functionality. Comparison of tool-like compounds with lead-like or drug-like compounds shows that as many or more compounds are filtered out by the presence of undesired chemical functionality as by the presence of undesirable physicochemical properties.

Rules and filters: do exceptions exist? Off course they do. Common sense is required. There is a natural priority order in drug discovery decision making. Clinical information trumps all. Next in importance is high quality experimental evidence, e.g., in vivo animal experiments. Rules and filters come into play when clinical and experimental data is lacking.

4.4 Structure Verification on the Original Sample

When a compound is reproducibly active in an HTS, confirmation of the structure is one of the very first activities. Is the identity of the original sample correct and what is the purity? Assurance as to chemistry identity is an evolving process. In the early 1990s, compound structures purchased from eastern European sources were sometimes fraudulently mislabeled. Today the situation is much improved but the purchase of screening compounds is still very much a "buyer beware" situation. It is common for organizations purchasing screening compounds to run their own analytical quality checks until they reach sufficient experience and confidence to trust the vendor's analytical quality checks. Assuming that the original sample of an HTS hit is the correct structure and is of satisfactory purity, more still needs to be done. History teaches that a presumed active sample needs to be freshly resynthesized and retested (often in a more careful screen than the primary HTS or HTS retest). Why? The sad fact is that somewhere between 10% and 50% of resynthesized samples prove inactive on retest. Often the reason for why it is a chemistry or biology issue is not known. In this authors opinion the retest issue is likely not uniformly distributed among targets. Simply put, certain targets are more sensitive to small quantities of electrophilic impurities. The ligands for these targets are more likely to not retest active on resynthesis because the original sample contained a small quantity of reactive impurity. How could one identify these types of targets? The mostly academic medicinal chemistry literature is a clue. If there are reports of ligands with reactive "warhead" groups active against one's target or a related target one should be exceptionally cautious about the retest issue.

4.5 Activity Verification in a Resynthesized Sample

The need to resynthesize compounds poses a dilemma for the chemist. Will he/she pursue and resynthesize a singleton active? The probability of resynthesis wasted effort on a singleton alleged active is higher than when nascent SAR is seen on a cluster of structurally related compounds. An SAR pattern even on as few as five or six structurally related compounds increases the chance that the activity is real. There is a tension between library design chemistry space coverage which is better with singleton compounds on the one hand and with library design with small clusters which is less efficient in chemistry space coverage but more efficient in terms of defining the true positives including the step of compound resynthesis. The advantages of small clusters in the HTS triage stage are sufficiently large that small clusters have been adopted as the preferred chemistry space coverage mode at AstraZeneca (AZ) and in the NIH Roadmap compound library acquisition effort.

4.6 *Hit to Lead also Known as Closed Loop*

Chemistry optimization is no longer a monolithic process in many drug discovery organizations. Traditional single stage SAR optimization has been replaced by two steps: an early "hit to lead" aka "closed loop" stage followed by a later "lead optimization" stage that in many respects is not that much different from the traditional medicinal chemistry optimization. The goal in the early stage is to learn as efficiently as possible about the liabilities and attributes of a chemical series. Is the chemistry quality high enough for the series to be pursued in the more labor intensive, slower and more difficult lead optimization process?

The type of chemistry very much dictates whether chemistry efficient "hit to lead" is a viable option. A hit derived from a chemical series that was originally prepared by some type of automated synthesis is a good candidate for "hit to lead." By way of contrast, a hit from a legacy hand crafted compound or a natural products semi synthetic modification may very well not be candidates for early stage efficient chemistry synthesis. The errors in the early 1990s in combinatorial chemistry have left a bad taste in many organizations for the term "combinatorial" chemistry." So other terms like "parallel synthesis" or "automated synthesis" or "chemical technology" are used. Today's libraries of automated compounds in "hit to lead" are typically fairly small, perhaps 20–50 compounds. The quality is very high in terms of assurance as to compound identity and purity. The idea is to maximize the information content on each compound. Rapid iterations of compound synthesis and testing with feedback to the next cycle of synthesis is termed "closed loop" chemistry and is a goal in many organizations. Assuming appropriate chemistry, the synthesis cycle time per se may be very short. Just in time starting material orders; the multitude of work station like equipment available to facilitate efficient chemistry and the routine ability to separate and purify compounds by reverse phase HPLC all increase the synthesis speed. As chemistry cycle times shorten, data cycle times may become rate limiting. Project in vitro biology is seldom rate limiting but ADMET data can easily be rate limiting. It does no good to plan the next round of synthesis based on in vitro project biology data if critical ADMET data on drug-like properties is missing. Delays in ADMET data can easily be exacerbated because it is so common for all testing to be performed on compounds in DMSO stock solutions. Centralized compound management organizations preparing and dispensing compounds in DMSO stock solutions were originally designed for HTS type operations. For the normal HTS, compound cycle supply times were not critical. It is only in recent years that the compound management groups have had to deal with the very fast compound cycle times required to support fast data cycle times in "hit to lead."

4.7 Multiple vs Single Chemical Series in "Hit to Lead"

Having multiple chemical series for the same biology approach is always an advantage in "hit to lead." There is the obvious advantage of multiple series giving more opportunity for success. However, there is another less obvious advantage having to do with people behavior. When things go wrong, as they frequently do, working on a single series is very dangerous. Unwilling to accept failure, medicinal chemists continue on a single series far longer than a more rational (less emotional) analysis would justify. Having a choice on switching to another series makes a change in chemistry much easier from a personal (and organizational) perspective. This is an important enough factor that some organizations will not enter "hit to lead" if there is only one chemical series. Furthermore, multiple chemical series set the stage for important input from computational chemists which can lead to opportunities such as "scaffold hopping" in order to unlock additional chemical space.

4.8 Profiling is Critical in "Hit to Lead"

Profiling of compounds is critical in "hit to lead." Every chemical series has problems. The goal in "hit to lead" is to find out what the problems are, to find out how many problems there are, and to assess how much progress can be made against those problems. If the process is successful the chemical series can be handed off to "lead optimization." Depending on the organization this next stage of chemistry may or may not be carried out by the same chemistry group or by a different chemistry group. Universally, organizations have set up exit criteria for leaving "hit to lead" or entry criteria for starting "lead optimization." The progression between the two stages entails a substantial increase in time and cost and so is a natural point for project planning and the application of go/no go decision criteria. The profiling in "hit to lead" should ideally cover all the major global issues in project biology and ADMET. The listing below is only a generalization:

1. Potency against primary project biology target
2. Selectivity against undesired targets
3. Aqueous solubility
4. Intestinal permeability
5. Metabolic stability
6. Drug-drug interaction potential
7. Transporter screen (e.g., *MDR1* for a CNS approach)
8. Toxicity screen (e.g., HERG or equivalent especially for CNS approaches)
9. Some type of input on systemic drug levels

Some types of screens used in "hit to lead" are controversial and are used in some organizations but not in others. Examples of these are:

1. Serum protein binding
2. Cellular cytotoxicity

The challenge for the drug discovery organization is how to handle the resource issues for multiple screening data feedback. Often a combination of experimental screens and computational prediction approaches will be used. Rapid data feedback to the medicinal chemist is essential, whether the data is experimental or computational. Data delayed is data with greatly reduced value.

4.9 How Many Problems Can Be Handled in Chemistry?

Realistically there is a limit as to how many serious problems in a chemistry series can be solved with a reasonable effort. In this author's opinion, optimism as to success declines markedly if there are more than three serious problems. What might a typical three problem profile look like:

1. Not potent enough (not selective enough) problem since potency and selectivity are often linked
2. Aqueous solubility
3. Metabolic stability

So the goal with this type of profile would be to see if progress could be made toward fixing these problems while monitoring those areas where no problem exists. This is important because while trying to fix one problem it is possible to inadvertently introduce a new problem.

4.10 Preformulation: Pharmaceutical Sciences in Early Discovery

Pharmaceutical sciences formulation technology in a technical sense can be used in early discovery including "hit to lead" to allow testing of compounds with poor physicochemical properties. Whether this intervention often called "preformulation technology" is a good idea is highly controversial. The propreformulation viewpoint is that without formulation assistance promising chemical series might be excluded from testing based solely on poor properties. Most of the time the issue is very poor aqueous solubility. The counter argument is that maybe another chemistry series should be looked at or the chemistry series should be dropped if the series has such poor properties that preformulation work is critical to get testing done. Regardless of a person's position on preformulation, nothing should be done in a pharmaceutical sciences sense in early discovery that would detract from the medicinal chemist's mission. The much preferred method of solving poor physico-chemical property problems is in chemistry. Change the structure and fix the problem. Unfortunately this is easier said than done. Solubility SAR tends to be blunt and the success rate at Pfizer at fixing solubility problems in chemistry was around 50%. Nevertheless the chemistry fix is the preferred choice. The formulation fix is the backup choice if chemistry efforts fail. Amorphous compounds predominate in early discovery and amorphous compound aqueous solubility can easily be 100 times higher than when the compound is crystallized. All parties need to be educated and informed so that, for example, seemingly acceptable solubility levels on amorphous materials are not misinterpreted by chemistry as indicating acceptable solubility. The bottom line is that involvement of pharmaceutical sciences in early discovery should be very carefully studied and debated in the context of organizational and people issues peculiar to the company.

4.11 Activity SAR Patterns

Progressing through "hit to lead" should provide evidence of true SAR. To be avoided are the extremes of flat muddy SAR (previously discussed) and the singleton active. Promising "hit to lead" series show good ligand efficiency. Across all targets this would be about 0.3 kcal mol^{-1} binding energy per heavy atom. Roughly this translates at a 10 nm target affinity to a MWT below 500 increasing the probability of oral absorption. Ligand efficiency varies by target class. For details the reader is referred to the specialist literature. A high ligand efficiency in a compound scaffold or core is a very promising sign. Experience at Abbott in a decade of fragment screening by NMR indicates that high ligand efficiency in a scaffold or core is a good harbinger of eventual success in series optimization. Rigorous use of ligand efficiency prevents a chemistry series from falling into the "hydrophobic trap," i.e., potency improvements entirely due to lipophilic functionality. By the time the in vitro activity is acceptable, e.g., 10 nm the lipophilicity log P is so far above the RO5 limit of five that solubility is terrible and oral absorption is near impossible.

5 Hit to Lead, Exit Criteria to Lead Optimization

Virtually every competent drug discovery organization has criteria for the transition between "hit to lead" and "lead optimization." Virtually nobody publishes these criteria because the organizations consider them proprietary. By virtue of very fast note taking and a combined effort, a Pfizer colleague and I were able to capture the "hit to lead" exit criteria from an oral presentation by a medicinal chemist from the AZ organization. The criteria are as they stood in about year 2000. I do not know if these represent current criteria from AZ. My sense is that the AZ organizations criteria lies towards the upper end of strictness range.

AstraZeneca (year 2000) "Generic lead target profile" for progressing HTS hits to leads:

Potency	100 nM
Rat hepatocyte intrinsic clearance	<14 μL $min^{-1}10^6$ cells
Human microsome intrinsic clearance	<23 μL min^{-1} mg^{-1}
Rat IV clearance	<35 mL min^{-1} kg^{-1}
	Volume >0.5 L kg^{-1}
	$T_{1/2}$ >0.5 h
Rat PO bioavailability	$>10\%$
Plasma protein binding	$<99.5\%$
Solubility	>10 μg mL^{-1}
CLogP	<3
LogD	<3
Mol Wt	<450
P450 inhibition IC50	>10 μM for five major isozymes
HERG screening	Early toxicity in vitro screens
Clear SAR around potential lead	
Selectivity – use PanLabs/Cerep batteries	
Structure must provide patent opportunities	
Need in vivo biological validation	

6 Conclusion

This chapter is the first chapter in this book and serves as an introduction The first section of this chapter starts with a discussion of "what is a medicinal chemist?" In this chapter the reader will notice far more attention to the "people" aspect of the discipline of medicinal chemistry than will be found in later chapters. This is by design. When expert practitioners in the field discuss success in medicinal chemistry the focus is usually very much on the person rather than on the technology. We admire the persistent "drug hunters." We remember the individuals who do not give up easily. We appreciate those individuals with the interpersonal skills that facilitate an effective therapeutic project team. We admire those medicinal chemists who think out of the box and come up with the insights that transform a program. In this introduction I have tried to give some credit to this "people" factor.

This introductory chapter is in some sense forward looking. Some discovery approaches such as phenotypic screening receive more attention in this chapter than they do across the discovery approaches of pharmaceutical and biotech companies. Drug discovery approaches are becoming increasingly fragmented. To cover the entire range of approaches is beyond the scope of an introductory chapter. The coverage on phenotypic screening reflects my own personal interest as well as my belief that this approach will grow rapidly in the next 5 years.

References

1. Gladwell M (2005) Blink: the power of thinking without thinking. little, Brown, New York, NY
2. Oprea TI, Davis AM, Teague SJ, Leeson PD (2001) J Chem Inf Comput Sci 41:1308
3. Wenlock MC, Austin RP, Barton P, Davis AM, Leeson PD (2003) J Med Chem 46:1250
4. Proudfoot JR (2005) Bioorg Med Chem Lett 15:1087
5. Leeson PD, Springthorpe B (2007) Nat Rev Drug Discov 6:881
6. Sams-Dodd F (2005) Drug Discov Today 10:139
7. Hopkins AL, Mason JS, Overington JP (2006) Curr Opin Struct Biol 16:127
8. Ho RL, Lieu CA (2008) Drugs RD 9:203
9. Paolini GV, Shapland RHB, van Hoorn WP, Mason JS, Hopkins AL (2006) Nat Biotechnol 24:805
10. Ertl P, Jelfs S, Muehlbacher J, Schuffenhauer A, Selzer P (2006) J Med Chem 49:4568
11. Hert J, Keiser MJ, Irwin JJ, Oprea TI, Shoichet BK (2008) J Chem Inf Model 48:755
12. Brown D (2007) Drug Discov Today 12:1007
13. Horspool KR, Lipinski CA (2003) Drug Deliv Technol 3:34
14. Congreve M, Carr R, Murray C, Jhoti H (2003) Drug Discov Today 8:876
15. Lewell XQ, Judd D, Watson S, Hann M (1998) J Chem Inf Comput Sci 38:511
16. Crisman TJ, Parker CN, Jenkins JL, Scheiber J, Thoma M, Kang ZB, Kim R, Bender A, Nettles JH, Davies JW, Glick M (2007) J Chem Inf Model 47:1319
17. Al-Lazikani B, Gaulton A, Paolini G, Lanfear J, Overington J, Hopkins A (2007) Chem Biol 3:804
18. Coan KED, Shoichet BK (2008) J Am Chem Soc 130:9606
19. Huth JR, Mendoza R, Olejniczak ET, Johnson RW, Cothron DA, Liu Y, Lerner CG, Chen J, Hajduk PJ (2005) J Am Chem Soc 127:217
20. Haggarty SJ, Schreiber SL (2007) Chem Biol 1:299
21. Verheij HJ (2006) Mol Divers 10:377
22. Hopkins AL, Groom CR, Alex A (2004) Drug Discov Today 9:430
23. Schneider N, Jaeckels C, Andres C, Hutter MC (2008) J Chem Inf Model 48:613

Top Med Chem (2009) 5: 25–83
DOI: 10.1007/7355_2009_6

Published online: 23 July 2009

High Throughput Screening in the Twenty-First Century

Jeff W. Paslay, John E. Morin, and Richard K. Harrison

Abstract High throughput screening (HTS) has become the cornerstone of lead identification for small molecules in drug discovery during the last quarter century. The evolution of the sciences and technologies that have evolved as the foundation of modern HTS campaigns are complex and require multidisciplinary interactions. Innovations in integrated automated systems, reagent systems enabled by molecular biology, computational capabilities, and visualization tools have converged to provide sophisticated tools to HTS practitioners. The success of HTS in an organization does not rest solely with those performing HTS but is critically dependent on the interactions of biology and chemistry members of the multidisciplinary teams throughout the early discovery process. Thus a basic knowledge and understanding of the components and processes of HTS is a necessary requirement for effective communication in planning, executing, and analyzing an HTS campaign. This chapter addresses the key components of HTS campaigns, common approaches, and related issues that should be understood by those engaged in small molecule drug discovery.

Keywords Assay development, Automation, Data management, High throughput screening, Statistics

---

J.W. Paslay(✉)
302 Old Barn Circle, Phoenixville, PA 19460, USA
e-mail: paslayj@wyeth.com,

J.E. Morin
3267 Street Road, Doylestown, PA 18902, USA
e-mail: morinj@wyeth.com

R.K. Harrison
1315 New Virginia Road, Downingtown, PA 19355, USA
e-mail: harrisr6@wyeth.com

Contents

1 Introduction

Discovering a drug and completing the subsequent phases of clinical development is hard. The long R&D process is complex and highly regulated with ultimate success dependent upon the interaction of multidisciplinary teams of biologists, pharmacologists, medicinal chemists, toxicologists, and clinicians. Today's process resulting in the launch of new small molecule therapeutic entities encompasses numerous steps from the initial idea of how to intervene in a particular disease through large scale clinical trials and each step has an attrition rate. Continually faced with pressures from consumers, stock holders, and management to improve efficiency and reduce costs associated with R&D, pharmaceutical companies have constantly evolved their approaches utilizing new drug discovery technologies.

In the late 1980s, many pharmaceutical companies began to adopt the concepts and technologies for high throughput screening (HTS) early in the discovery process to increase identification of molecules that would enable chemists eventually to develop drugs for entry into the clinical pipeline. During the early 1990s with an increasingly competitive market coupled to a highly charged political environment, meeting the challenge of maintaining double digit revenue growth became a number one priority of the entire pharmaceutical industry. One study commissioned by ten pharmaceutical companies evaluated all aspects of the discovery process to determine how to restructure programs to develop the high performance required for future success. The results focused on broad strategies, processes, technologies, and organizational and informatics capabilities. The implementation of a strong infrastructure to support HTS for lead identification was one of the six critical process-related best practices recommended for success [1].

Driven by developments in computational capacity, laboratory scale automation, combinatorial chemistry, and evolving targets from knowledge of the human genome, HTS has become an integral element in most lead identification strategies for small molecules, natural products, and protein therapeutics. Investment in the last 10–15 years has built upon early HTS capabilities and now provides an industrial scale approach implemented in most pharmaceutical and biotechnology companies. Instrumentation and automation development has continuously increased the speed and capacity to allow modern systems to screen hundreds of thousands of compounds per day. The size of sample libraries has grown from tens of thousands to millions as parallel synthesis technologies evolved and the capacity of HTS systems grew. Data management applications and storage capacity have enabled the HTS practitioner to review, analyze, and annotate millions of data records. Most importantly, a community of HTS scientists has evolved with the necessary perspective, skills and knowledge to utilize effectively the technology at hand to impact early drug discovery. The evolution of HTS has provided a process leading to marketed products and populated pipelines at most companies. The scientific disciplines and technologies involved in HTS have begun to migrate upstream in target identification and downstream in support of lead optimization and ADMET evaluations. Recently companies have begun to discuss and present their efforts and successes in HTS over the last 20–25 years [2], and it is evident that HTS has become and will remain a cornerstone of drug discovery for the foreseeable future.

In this chapter, our goal is to provide a working knowledge of HTS with enough detail to enable and enrich the communication between chemists or biologists and their HTS colleagues. The discussions of concepts, processes, and technical details are interspersed with some of our approaches at Wyeth used only as examples.

2 Useful Definitions

As with most scientific disciplines, HTS is understood and discussed only with a thorough comprehension of the descriptive terms used by HTS practitioners. A brief list of relevant terms and definitions which will aid the newcomer to the field is provided in the Appendix. For a more comprehensive list of terms and definitions, which is constantly being expanded, the reader may find the Society of Biomolecular Sciences website helpful [3].

3 Conducting an HTS

3.1 Automation

HTS is most simply defined in terms of throughput. The use of automation to achieve high throughput is a natural, but not inevitable, consequence in the quest for speed. Several additional factors have contributed to the adoption of large, complex, automated assay systems by most pharmaceutical companies engaged in drug discovery. Automation provides increased robustness based on repeatable processes, reduced staff involvement in performing the screen to allow monitoring of performance and data analysis, and increased quality due to similar handling of each assay plate with sophisticated scheduling software. Even though automation does offer significant opportunities to improve the HTS process, it can be seductive and should not be pursued as an end in itself. In the words of a past President of the Society for Biomolecular Sciences, "It is irrelevant if the plate density is lower or higher, the process is fully automated or purely conducted by hand...;" the most important criteria for successful HTS are robust and predictive biological assays, rich and diverse screening libraries properly maintained, and standard operating procedures with adequate quality control [4].

3.1.1 Structural Organization

The historical evolution of HTS groups from small groups of assay specialists serving therapeutically focused departments and from engineering support groups charged with evaluation of new technologies naturally favored decentralized HTS operations in many companies. The advantages of decentralized HTS operations largely derive from close proximity to their client groups and responsibility to support only a limited number of key assay technologies that enable specialization of procedures and equipment and development of world class expertise. Reliance on decentralized HTS operations can encourage failure to communicate and coordinate activities across groups within a company, in extreme cases leading to creation of competing "centers of excellence." Another disadvantage of decentralized organizations can be vulnerability to the abrupt shifts in focus that are sometimes necessitated by strategic considerations. If personnel and equipment are ideally specialized to serve the HTS needs of a particular client group, they may not be readily redeployed.

As companies strove to achieve cost effective policies across discovery research organizations, centralization became an attractive alternative to decentralized HTS operations. Centralized HTS operations afford tight management control of resource allocation and project prioritization. Central HTS groups generally command larger budgets that can be strategically managed to exert maximum leverage with vendor companies supplying automation, reagents, and consumable supplies. The most important advantage of centralized HTS organizations is arguably the ease with which uniform standard operating procedures and quality controls can be developed and maintained across all platforms and HTS campaigns. In addition, the compound management function is most often a centralized core function and it can be extremely convenient to co-localize the compound management and HTS functions. The major disadvantage of a centralized HTS organization is the need to transfer knowledge and trust from the client groups responsible for target identification and assay validation to the screening group, often across distances that discourage personal interaction. The desire to avoid the necessity to communicate across several time zones is still a cogent argument for co-localization of decentralized HTS groups with the client groups that they serve.

3.1.2 Platforms

A comprehensive listing of all the vendors that offer HTS instrumentation and platforms is beyond the scope of this chapter, but most vendors maintain informative websites and there are three professional organizations that disseminate useful information about automation platforms for HTS on the world wide web, the Society for Biomolecular Sciences (www.sbsonline.com), the Association for Laboratory Automation (www.labautomation.org), and the Laboratory Robotics Interest Group (www.lab-robotics.org). The latter maintains an online forum where vendors and experienced users often provide immediate and useful guidance.

HTS fundamentally requires only four component instrument: precision pipettors that can quickly dispense many different test compounds from the screening library in parallel to separate wells on a plate, precision bulk dispensers that can quickly dispense the same reagent solution to all the wells on a plate, incubators to maintain reaction conditions and optical plate readers that can quickly determine the readout from the assay. In extreme situations, highly motivated scientists can maintain sufficient control of the sequence and timing of complex assays to process 40–50 assay plates a day using manual instruments that require the operator to place the plate, activate the instrument, and remove the plate, moving the plates from one instrument to the other, in sequence, while manually tracking the flow of data from the plate reader and annotating the data stream with the samples tested. While barely qualifying as HTS, a manual operation can complete hundreds of plates over a period of months and process a substantial corporate screening library. It is generally agreed, however, that the productivity of the exceptional scientist capable of sustaining a manual screening campaign can be greatly enhanced by implementing some automation.

Most instrument vendors offer optional plate stackers that will accept 50–100 plates from an input stack and process them sequentially through the instrument and collect the processed plates in an output stack. These instruments generally perform one function on a moderate number of plates, commonly either liquid handling or plate reading. Some semi-automated work stations combine liquid handing and plate reading functions in one stand-alone device. These more elaborate stack fed combination instruments are often referred to as semi-automated workstations and they generally include a robotic control application with a graphics user interface and data management capabilities for visualizing and exporting output files from the plate reader.

The next step in complexity and functionality in HTS automation can be defined as the workcell concept, in which several different instruments serving the four main functions are integrated into a relatively dense array on a fixed framework that includes some robotic mechanism for transporting plates from one instrument to another. Workcells may use stacker fed instruments or robot accessible instruments, but in either case a key distinction is that the robotic control application is now responsible for maintaining and tracking the movement of assay plates through all the steps of the assay according to a well-defined schedule. This use of the term workcell, while common usage, is not well defined, and the scale of these automated systems varies, but they most commonly consist of one pipettor, several dispensers, one incubator, one plate reader and one robotic arm in a containment cabinet that places everything within reach of the one arm. Typically, the workcell vendor is responsible for subcontracting all other components and assembling and servicing the system. The functional capacity of a workcell may not exceed what can be achieved with semiautomated workstations, but they afford more freedom to the operator for other tasks such as reagent preparation, data management and development of protocols for upcoming projects. The compact size, modular design and potential for vendor supported standardization across sites would seem to make workcells particularly apt for de-centralized HTS operations.

The most complex automated systems are used almost exclusively by centralized HTS operations in large pharmaceutical companies and are referred to as ultra HTS (uHTS) platforms. They typically consist of the same four functional instruments, but have the capacity to process several hundred plates per extended workday. They often incorporate a modular design philosophy with multiple duplicate instruments for enhanced capacity that offer some functional redundancy. The mechanism for moving plates from one instrument module to another is often, but not always, a continuous track-way that resembles an industrial assembly line rather than the robotic arm typically used in a workcell system [5–8].

3.1.3 Art of the Possible

Use of uHTS devices incurs significant capital and operating expense plus requires extended operator training and some form of resident engineering and facilities support. The assay protocols required to achieve these throughputs impose some limits on the flexibility of the systems. The cycle time for each individual operation on each of the pipettors, dispensers, or plate readers must usually be limited to no more than 1 min per plate in order to sustain the desired throughputs. This effectively limits most assays to end-product determinations and imposes limits on the number of reagent addition and incubation steps and on the intermediate incubation intervals so as to maintain the largest possible continuous batch size of plates for the most cost effective operation. The extended run times for large plate batches impose more stringent requirements for stability of reagents and most uHTS devices include some accommodation for onboard refrigerated reagent storage. In order to maintain flexibility to utilize different assay readouts, most uHTS platforms include multiple, multimode plate readers capable of reading absorbance, luminescence, chemiluminescence, time resolved and prompt fluorescence. There are some readouts, however, such as fast response prompt fluorescence, that can only be detected by specialized readers such as the FLIPR, and require a specialized uHTS platform.

Other readouts, such as high content imaging or enzyme kinetic rate determinations, require extended detection intervals in the plate reader and are best automated on a dedicated workcell or semiautomated workstation platform.

Despite the costs associated with implementation of automation, nearly all drug discovery programs utilize some form of automation. Even organizations that favor hit identification methods other than HTS use automation for target validation assays, reagent preparation, protein crystallization, and NMR spectroscopy. The most compelling reason for implementing automated HTS systems (workcells and uHTS platforms) may be the conclusion reached by the pioneering scientist who commissioned one of the first uHTS systems in 1997. The precise control of variation in environmental conditions, reagent storage, dispensing, and timing that was achieved with the first uHTS system significantly enhanced the overall quality of the assay results compared to the semiautomated workstations then in widespread use. He concluded that enhanced standardization of process and quality control might be more important benefits in the long run than enhanced capacity [5].

3.2 HTS Process

Successful management of a central HTS operation requires a discipline of process control that extends beyond the automation systems. The progression of HTS projects must be carefully guided and tracked through several milestones designed to ensure that a steady and predictable stream of robust assays are available for transfer to the automation platforms and that sufficient bulk reagents and consumable supplies will arrive in time to support the HTS campaigns.

3.2.1 Phases of HTS

It would be very rare to find a target that was conceived of, biologically validated, and pharmacologically confirmed that originated in an HTS laboratory. Generally, target identification and validation are the responsibility of a therapeutically focused research area or department. Each pharmaceutical company has their own unique process to cover the transition of a target from validation through to the confirmed hits delivered by an HTS campaign. Despite these variations, one can identify five common functions:

- Transfer of an assay from therapeutically focused area to HTS
- HTS assay development and validation
- Robotic assay adaptation and validation
- HTS campaign
- Hit confirmation

We will examine each of these phases in greater detail, outlining the deliverables associated with each phase.

3.2.1.1 Transfer of an Assay from Therapeutically Focused Area to HTS

Assays which originate on a bench top in a research lab are generally not suitable for HTS. They are developed to help with biological validation or characterization of tool compounds. Thoughts about the number of reagent additions, length of incubations, number of transfers, miniaturization, signal to background, and cost of reagents are not the primary driver at this point. Yet these are precisely the details which are important to run a successful HTS campaign. Therefore, most organizations have a group of scientists who are responsible for shepherding a target from the research group to the HTS group. The most successful organizations will adopt a formal process that requires projects likely to benefit from an HTS campaign to consult early and often with the HTS group on the choice of assay format to be used for the HTS campaign. The best use of the HTS resource is usually a balance between the cost and logistics of HTS and the fidelity of the biological response in the automated assay to the "gold standard" laboratory assay.

The first step in the process involves the development of a suitable HTS assay format. A more detailed discussion of the formats available for HTS by target class are outlined later in this chapter. This section will focus on some of the practical considerations to determine the proper assay technology for an HTS campaign.

The first question to be answered is can the assay technology output be detected and properly quantitated on the HTS platforms? While this is a straightforward and seemingly obvious point, it is quite often overlooked. In an extreme case, a radiolabeled ligand binding assay was initially developed by a research laboratory for an HTS platform that could only read fluorescence. Most HTS platforms only read fluorescence or luminescence, and this will determine the range of options for assay format. There are numerous choices for format, but there are some general guidelines to follow. Simple mix and read, or homogeneous formats, are generally better than heterogeneous formats that require washing or some other separation of product from reactant for detection. Cost of reagents is always a factor. This includes biological reagents, consumable labware (plates, pipette tips, etc.), etc. At present, an acceptable average cost per well for these components would be 5–10 cents per well.

The range of possible assay formats varies between organizations, but a list of the parameters which should be addressed include:

1. Optimizations of S/B, EC_{50}/IC_{50}, Z'
2. DMSO tolerance
3. Optimization of buffer components
4. pH optimization
5. Reducing agents
6. Monovalent/divalent cations
7. Metal chelators
8. Enzyme "stabilizers"
9. Base buffer
10. Detergents

11. Cell density
12. Reagent concentrations and volumes
13. Liquid handling steps
14. Incubation steps
15. Substrate/ligand/inhibitor specificity at optimized conditions
16. Stability up to 5 h
17. Indication of the potential hit rate using a standard set of compounds
18. Reproducibility of S/B, EC_{50}/IC_{50}, %CV:
    (a) Within three experiments
    (b) Within >three plates

In most cases, in consultation with the HTS group, the research area laboratory will develop a benchtop assay that is at least compatible with the HTS format of choice for their target. This tends to facilitate project transitions and provides a tool that the research area laboratory will use later to follow up hits and develop SAR. In other cases, the HTS assay development group will assume all responsibility for assay development. The formality of the transfer of the project from the research area to the HTS group varies between organizations, but the outcomes are quite similar. All of the details of the prototype assay are reviewed by both teams, and, where applicable, reagents, protocols, and even plates or pipette tips are exchanged.

HTS assay transfers often occur across separate sites and require excellent communication and close collaboration. This milestone is often marked with a meeting and documentation generated that records the protocol and the status of the assay results in the hands of the research laboratory.

3.2.1.2 HTS Assay Development and Validation

Assay development by the HTS assay development group begins with a reconfirmation of the research laboratory assay in the HTS assay development laboratory. This is usually accomplished by reproducing time courses for the reaction, reproduction of standard assay conditions, and validation using the empirically derived IC_{50} or EC_{50} of a known target modulator where applicable. The HTS assay development group then further develops their assay protocol using semiautomated workstations and plate readers similar to those on the uHTS systems and validates any modifications to the protocol or materials that may be necessary to achieve robust, large-scale, batch operations. Wherever possible, materials and methods must be standardized from project to project to shorten timelines, build expertise, and afford economies of scale in purchasing reagents, labware, and cell culture supplies. Nevertheless, new assay formats are continually introduced to the market, so a balance must be achieved between the familiar and the novel. The final assay development exercise is a test of the small, diagnostic compound library, like LOPAC (Library of Pharmacologically Active Compounds), commercially available from Sigma. This test ensures that the hit rate in the assay as configured will not overwhelm the compound sample management group's ability to pick hits for confirmation analysis and also may be used to identify a suitable reference compound for quality control testing, if a well characterized reference standard with a known mechanism of action is not available. Following validation of the assay and reagents by the core assay development group, the assay is finally transferred once more into the hands of the HTS group who will further adapt the protocol as necessary to achieve acceptable performance on a fully automated HTS platform.

3.2.1.3 Robotic Assay Adaptation and Validation

The remaining HTS process is divided into four phases: HTS adaptation, HTS validation, HTS online, and HTS confirmation. Altogether the four phases require 1–4 months to complete, depending on the assay format. The average screen is completed in 2.5 months. The first phase of the process is the development of an automation protocol that emulates the semiautomated procedure developed by the assay development group and delivers similar results from controls and/or reference compounds. This HTS adaptation phase will be more or less successful and take more or less time depending on the assay format and statistical robustness. Well-behaved, homogeneous format assays can often be adapted to automation in a couple of weeks while cell-based assays with extremely sensitive cell types sometimes require months of effort to develop a robust protocol. As soon as the fully automated assay protocol yields results similar to those achieved with semiautomated instruments in the assay development group, the HTS group will begin assay validation, i.e., scaling up the batch size to achieve at least 40, 384-well plates per batch. As the methods are adapted to high throughput technology, any changes in the protocol are noted and reviewed by the therapeutic area. As soon as large batch operations yield reproducible results, the validation process is expanded to include replicate tests with LOPAC plates. The LOPAC tests at this point serve a different purpose than those in assay development. Multiple replicate plates are tested over several days to evaluate correlation of results from plate to plate and from batch to batch. An uninterrupted supply of validated bulk reagents and/or cells is crucial from this point forward. Statistical analyses of the results obtained from controls and/or reference compounds are more robust when many replicates from several separate test events are processed. The results must demonstrate low variability from well to well and plate to plate from day to day or it will be impossible to distinguish biologically active samples from statistical fluctuations. If results from available reference standard compounds are acceptable and variability is acceptably low and replicate sets of LOPAC results show strong correlation, the screen is said to pass HTS validation and goes online. This milestone is also marked with a meeting that includes the therapeutic area project team, the core assay development group and the core HTS group and is documented with presentation slides including the results of the LOPAC library and the final HTS protocol as well as the first uploads of results data to the corporate repository.

3.2.1.4 HTS Campaign

While online, most HTS campaigns will test more than half a million compounds once at a single concentration of 10 μM in either 384- or 1,536-well plates. In collaboration with Wyeth biometrics specialists, an activity threshold of three interquartile range standard deviations (IQR SD) from the population median of the entire HTS screening library has been adopted for hit selection. Results are first normalized on a per plate basis, often using results from control wells located on each plate. This practice typically generates hit rates between 0.5% and 3% of the library. The hits are "cherry-picked" from the automated compound store and replated for confirmation testing. Other successful HTS operations use other methods to select hits for follow-up. Often an arbitrary limit is set on the number of hits that will be retested based on resources available for "cherry-picking" and confirmation testing and the selections is then based on apparent potency and/or chemical structure criteria such as scaffold diversity, cluster analysis, pharmacophore modeling, etcetera. The statistical analysis approach casts the widest net and does not discriminate against novel singletons. It often rescues compounds of modest initial potency that are ultimately selected as better starting points for lead development than more potent compounds and compounds that resemble known pharmacophores.

3.2.1.5 Hit Confirmation

The confirmation phase includes a titration analysis at multiple concentrations in triplicate as well as an interference screen, designed to identify compounds that show apparent activity through nonspecific interference with the assay reagents. The interference screen is an additional control and is performed under conditions that are nearly identical to those used in the primary screen to help insure that interfering compounds behave in a similar fashion in both the target and interference assay. The simplest interference screens simply add a carefully matched amount of previously prepared product to the standard assay in the presence of a candidate hit to rule out the possibility that the compound alters the readout to appear active through some trivial, non-mechanistically relevant effect. Screening hits identified in the online phase must demonstrate at least as much activity against the target and not show activity in the interference screen in order to be designated as confirmed hits [9].

At this point, a campaign review can be presented to the therapeutic area project team, including a list of hits with confirmed biological activity and a thorough documentation of all the quality control results across the entire campaign. All the biological data generated by HTS is released to the corporate database. SpotFire visualization of dose titration curves is presented to help group the confirmed hits into several different classes according to potency. The project team is urged to select their best hits for chemical purity and integrity testing by an analytical chemistry group. The compound management group reformats these samples for purity and integrity testing and also provides additional sample for secondary testing by the therapeutic area.

These milestones and the meetings and documentation generated for them are a necessary aspect of project management and quality control for uHTS campaigns. While the details of the process vary widely across the industry, the use of automation to test a large collection of compounds and then subjecting the hits to increasingly rigorous follow-up testing is general.

3.2.1.6 Logistics

The scale of operations of uHTS facilities demands special attention to logistics in the design of the facility. Some uHTS facilities prosecute between 20 and 50 campaigns a year and routinely have several different campaigns ongoing concurrently. Ideally, the facility will incorporate its own freight delivery and waste pickup access points as well as adequate storage either within the facility or nearby to maintain stocks of bulk reagents, consumable labware and a spare parts inventory for the automation systems.

The demands for protein reagents and cells for HTS have already prompted several companies to adopt automated systems for protein purification [10, 11].

Traditionally, in order to achieve a specific cellular response with sufficient signal to background and reproducibility from day to day, HTS required the selection and continuous culture of stable, genetically transformed, cell lines. It could take months to create the stable transformed cell line for a particular target. The continuous culture and daily harvest of successive batches of these cells with different passage histories over the course of the HTS campaign often led to variability of the cellular response in the HTS assay. Over the past several years, it has become common practice to culture cells using automated bioreactors in advance of an HTS campaign to create a large stock of cryo-preserved cells stored in individual aliquots that are thawed as needed to provide enough cells for a batch of screening plates. The freshly thawed cells can be dispensed into plates and used immediately or after overnight incubation for some assays, while for others the cells must be dispensed initially into tissue culture flasks and incubated for several days before they can be plated and used for assay. Even when cryo-preserved cells must be transiently cultured in flasks to recover from cryo-preservation before use, when all the cells that enter screening have identical passage histories, the assay results from batch to batch tend to be more uniform. Many companies routinely use automated systems for cell culture and preparation of cells in plates. In addition to the expected benefit in quality control due to uniform procedure, automated cell culture systems are often reliable enough to enable off-shift and weekend activities that are often necessary to support full utilization of the HTS platforms [12, 13].

A recent strategy that builds on the widespread success with bulk cryo-preserved cells in HTS is the use of transiently transformed mammalian cell lines so that the time and effort associated with creation of stable transformed cell lines can be avoided. This technique has been demonstrated using conventional lipofection transfection technology [14]. An alternate transfection technology has been introduced that utilizes Baculovirus vectors that were originally developed for insect cells. Because Baculovirus vectors can carry mammalian genetic expression cassettes into mammalian cells but cannot replicate in them, the transduced cells retain the level of biohazard risk associated with the parent cell type. Because the cells are prepared in large batches and stored frozen until use, they provide a uniform response. An added benefit is that transient expression may enable the study of recombinant gene products that cannot be isolated in stable cell lines because they have a toxic effect or interfere with the cell cycle [15]. These practices tend to uncouple HTS from cell culture and simplify the logistics of screening.

3.3 Assay Development

Most drug discovery targets fit into one of the target families shown in Fig. 1. Included in the figure is the distribution of the market share by target family.

Each of the target families have been reviewed in great detail elsewhere, and the reader will be guided to the appropriate references. This review will quickly survey the assay technologies most often employed in HTS facilities. First, a brief review of the common assay techniques used across the target families will be presented. Next, specific assay formats for the major target families will be presented. Each family has its own special assay needs and therefore specific requirements for effective HTS assay formats. However, a few generalizations concerning format can be found throughout the families. For all of the assays reviewed here the, format

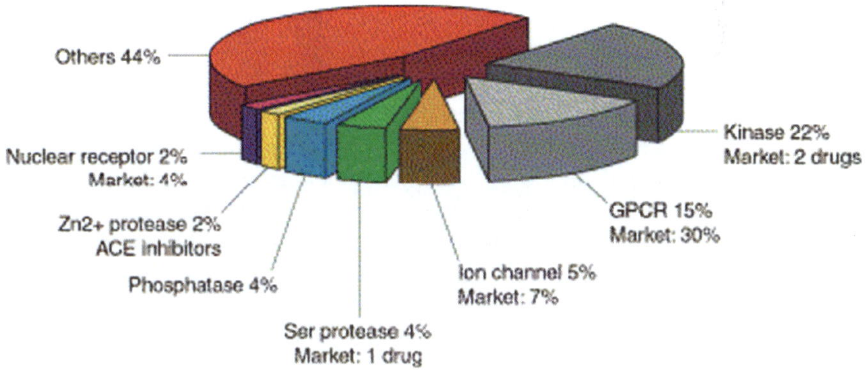

Fig. 1 The distribution of current drugs by target class. Included in the figure is the current percent of market share

must be homogeneous. HTS assays which involve separation steps or washing severely decrease throughput and efficiency. They must all be able to run in higher density formats, 384- or 1,536-well plates. Each must be able to be run in small volumes, preferably 10μL or less, to both decrease reagent use and waste, and keep costs low. Endpoint assays are preferred to kinetic assays, but it must be shown that the reaction is linear over the time course being investigated.

Prior to a detailed analysis of target specific assays, it would be prudent to review some of the homogeneous technologies that have recently become commercially available. The last 10 years have seen a substantial increase in the variety of techniques available for measuring biochemical activity. A few of the more popular techniques will be reviewed here.

3.3.1 Assay Formats

3.3.1.1 Fluorescence Polarization

Fluorescence polarization (FP) is a powerful fluorescence-based technique for the study of biomolecular interactions in aqueous solution. Small molecules rotate quickly, large molecules rotate slowly. By using a fluorescent dye to label a small molecule, its binding to another molecule of equal or greater size can be monitored through its decreased speed of rotation. This is the basis of FP. Assays are designed to change the size of a fluorescently tagged molecule, hence changing its rotation speed.

First described in 1926 by Perrin [16], the theory was greatly expanded by Weber [17], who developed the first instrumentation for the measurement of FP. Dandliker [18] expanded FP into biological systems such as antigen-antibody reactions and hormone–receptor interactions. Jolley [19] developed FP into a commercial system for monitoring of therapeutic drug levels and the detection of drugs of abuse in human body fluids.

In order to understand what we measure in FP it is important to understand first the nature of light absorption and emission. If we represent the molecules as dipoles, the excitation dipole is the direction in which the molecule prefers to absorb light. The emission dipole is the direction in which a molecule prefers to emit light. We assume (for the sake of simplicity) that these directions are parallel. If all the molecules are aligned with their excitation dipoles aligned in the vertical plane and we shine vertical plane polarized light onto them, and we do not allow the molecules to move, then the observed polarization is 1 (commonly referred to as 1,000 mP). If all the emission dipoles are perpendicular to the polarized light the observed polarization will be 0. In reality molecules are not static, but are tumbling in solution. This leads to polarization values much lower than the theoretical limits.

FP assays can be developed for most of the target families. Examples of these assays are presented in the next section.

3.3.1.2 Homogeneous Time Resolved Fluorescence

Homogeneous Time Resolved Fluorescence (HTRF) (Cisbio International) is an assay based on the proximity of a lanthanide cryptate donor and a fluorescent acceptor molecule whose excitation wavelength overlaps that of the cryptate's emission. The utility of this technique is based on the time resolved fluorescence properties of lanthanides. Lanthanides are unique in the increased lifetime of their fluorescence decay relative to other atoms, so a delay in collection of the emission intensity removes the background from other fluorescent molecules. An example of the HTRF assay is a generic protein-protein interaction assay shown in Fig. 2.

When the proteins are in close proximity the Europium-cryptate emission can be absorbed by the acceptor (such as allophycocyanin [APC], or XL) which emits at a higher wavelength. When the two proteins are far apart, no fluorescence resonance energy transfer (FRET) occurs.

Europium cryptates are ideal detection chemistries due to their inherent stability and low background fluorescence. Variations on the HTRF scheme include the LANCE and Lanthascreen assay formats. Recently, a number of different lanthanides have been employed in HTS assays, among then Tb and Sm, each with its own corresponding acceptor. This has allowed for a broad number of assays formats. The use of time gating minimizes the emission signal from the free acceptor coupled with wavelength filters to remove the emission signal from the donor. The readout on these assays is based on the signal ratio of the donor and FRET emissions thus removing the impact of colored compounds which may decrease both signals, but the ratio will remain constant.

3.3.1.3 Bead-Based Assays

In the last decade a number of assays based on binding of substrates or products to beads have been launched. The beads are used as a means of separating or isolating one of the components in the assay mixture which is then able to give a report on the status of the substrates and or products. An example is the AlphaScreen (amplified luminescent proximity homogeneous assay) technique. In this assay, binding of molecules captured on the beads leads to an energy transfer from one bead to the other, ultimately producing a luminescent/fluorescent signal.

AlphaScreen assays utilize two types of bead, donor beads and acceptor beads. Donor beads contain a photosensitizer, phthalocyanine, which converts ambient oxygen to an excited form of O_2, singlet oxygen, upon illumination at 680 nm. Like other excited molecules, singlet oxygen has a limited lifetime prior to falling back to ground state. Within its 4-μs half-life, singlet oxygen can diffuse approximately 200 nm in solution. If an acceptor bead is within that proximity, energy is transferred from the singlet oxygen to thioxene derivatives within the acceptor bead, subsequently culminating in light production at 520–620 nm. In the absence of an acceptor bead, singlet oxygen falls to ground state and no signal is produced.

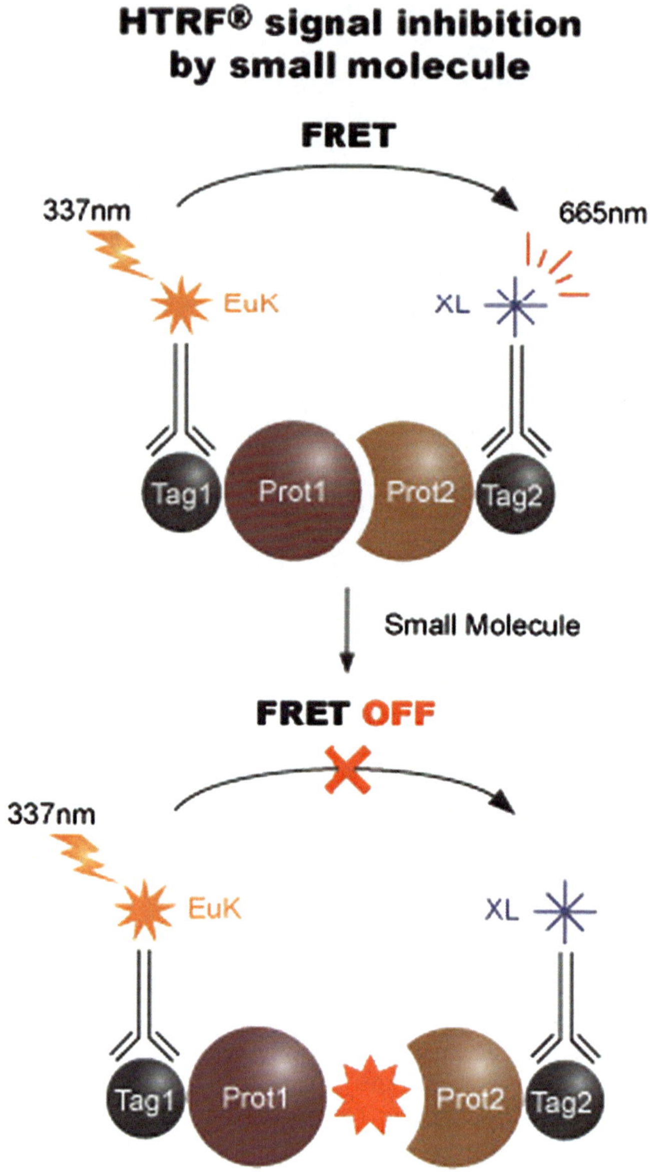

Fig. 2 A schematic representation of an HTRF assay for a protein–protein interaction. One protein is tagged with a fluorescent molecule whose emission spectra overlaps with the excitation of another fluorescent molecule. When they are in close proximity (*above*) the energy is transferred. When they diffuse apart (*below*) or are inhibited from coming together by a small molecule no FRET occurs

3.3.1.4 Scintillation Proximity Assays

Radioactive assays have fallen out of favor over the last decade, and few HTS assays are run using radioactive tracers. Safety, cost, and the need to dispose of large amounts of waste, along with the availability of nonradioactive formats have hastened their decrease. However, scintillation proximity assays (SPA) are still run in some HTS laboratories.

SPA is based on bringing a radioactive species in close proximity to a bead of solid scintillant. The technique relies on the specific capture of the substrate or product onto the bead so that the radioactivity can be measured without the need for separation.

3.3.2 Assay Formats by Target Class

3.3.2.1 Kinases

Several excellent reviews have been written over the last decade highlighting the many different kinase assay formats available and their application to specific enzymes [20–22]. This section will only briefly review the current formats used within the HTS environment. The reader section consult the above-mentioned references for greater detail of each of the formats.

Kinases are enzymes that place a phosphate group on a serine/threonine or a tyrosine residue of a protein or peptide. All kinase reactions use ATP as the phosphate source. Therefore there have been assays developed that monitor the loss or gain of the peptide/protein substrate (LANCE, ULight) [23], the loss of ATP (easylite luminescence kinaseGlo, Perkin Elmer) [20], or the gain of ADP (Transcreener TR-FRET) [24]. Many of these formats are applicable to cell based assays.

The format most widely used in the HTS environment is the FRET assay to measure the production of phosphorylated peptide product. The two most popular assays using this format are the LANCE and ULight assays (Perkin Elmer). Both assays employ an Eu-labeled antibody which binds to the phosphorylated product. The difference between the two formats is in the acceptor molecule (Fig. 3). Traditional LANCE used APC, which cannot be directly added to the substrate. In most cases the APC is bound via a biotin tagged on the N-terminal of the substrate and coupled through a streptavidin conjugated to the APC. In the Ulight assay, the Ulight acceptor is directly labeled to the substrate. The advantage of direct labeling this small acceptor is the elimination of the steric hindrance caused by the bulky biotin–streptavidin conjugate in the LANCE.

Another popular assay format for kinase assays is the Lanthascreen. This format is a variation on the LANCE assay, but employs Tb as the cryptate. In this format N-terminally fluorescently tagged peptide substrate (acceptor) is phosphorylated by the kinase. Next, a phophospecific antibody which is labeled with terbium binds specifically to the phosphorylated product, placing the donor and acceptor in close proximity, generating a signal [25].

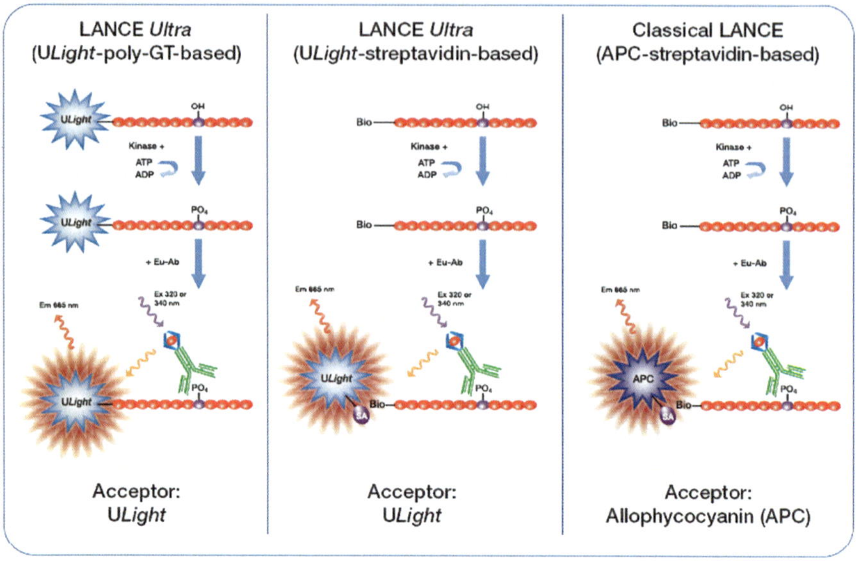

Fig. 3 A schematic representation of a LANCE ulight assay (*left*) and a traditional LANCE (*right*). See text for details

A more recently introduced format is the AlphaScreen assay. The assay principal behind this technology has previously been described above. In the kinase format a biotinylated peptide is bound to a streptavidin donor bead, and a phopshospecific antibody is bound to the acceptor bead. When the substrate is phosphorylated, the beads come in close proximity and a signal is generated. An example using the assay for the detection of inhibitors of serine kinases is presented by Von Leoprechting [26].

3.3.2.2 Proteases

Proteases are enzymes that break peptide bonds in proteins. As such they lend themselves to a variety of homogeneous assay techniques. Most employ labeling both ends of the substrate with a different tag, and looking for the appearance (disappearance) of the signal generated in the intact substrate (product). As an example, for a fluorescence quench assay, the N-terminal of a peptide is labeled with DNP and the C-terminal with MCA. As such, the peptide is fluorescently silent since the fluorescence from DNP is quenched by absorption by the MCA. Another very popular donor/acceptor pair is EDANS: 5-[(2-aminoethyl)amino] naphthalene-1-sulfonic acid and DABCYL: 4-(4-dimethylaminophenylazo)benzoic acid) (a sulfonyl derivative (DABSYL) [27]. Upon peptide cleavage, the two products diffuse, and due to a lack of proximity, the fluorescence increases.

Numerous variations on this theme exist [28, 29]. Assays utilizing FRET pairs are found throughout the literature [30], including homogeneous time resolved quenching [31]. One of the disadvantages of this technique is the inner-filter effect. At high substrate concentrations the fluorescence form a product can be absorbed by an intact substrate causing a decrease in fluorescence. Good assay protocols keep substrate concentrations below the concentration that causes this effect.

Another interesting protease assay technique involves FP [32]. A peptide is labeled with a fluorescent group at one end (usually near the C-terminal). The N-terminal is labeled with a biotin group. When the peptide is bound to the bead the FP signal is large. Protease catalyzed cleavage of the peptide releases the fluorescent tag, and the FP signal decreases. Variations on this theme include a FRET based FP assay. In this format a peptide is labeled on the N-terminal biotin and on the C-terminal with a FRET acceptor. When bound to an anti-biotin antibody the intact peptide produces a high FRET signal, and upon cleavage of the peptide, the fluorescent group is lost and the FRET decreases. Numerous other examples of protease assays exist in the literature.

3.3.2.3 Nuclear Receptors

Nuclear receptors (NRs) are ligand activated transcription factors. This class of targets includes the steroid and hormone receptors, glucose, lipid, and xenobiotic sensors. Most NRs exist in the cytoplasm of the cell, and when ligand binds, they translocate to the nucleus where they recruit transcriptional machinery and turn on gene transcription. Of the 50 NRs identified in the human genome, 5 of these are targets of current therapies. Examples include the oral insulin sensitizers Rozaglitazone, which are PPAR agonists.

Current assay technologies to measure agonists and antagonists of NRs range from simple fluorescent binding assays to FP assays. An example is the binding of fluorescently labeled dexamethasone to the glucocorticoid receptor [33]. In the absence of binding the ligand has a low polarization, but in the presence of receptor exhibits a high polarization. A more detailed assay involves the recruitment of a coactivator peptide via a FRET interaction. Such a pair is shown in Fig. 4. Cryptate-labeled anti-GST antibody indirectly labels the NR by binding to the GST tag. Binding of the agonist to the NR causes a conformational change that result in an increase in the affinity of the NR for a coactivator peptide. The close proximity of the fluorescently labeled coactivator peptide to the cryptate-labeled antibody causes an increase in the TR-FRET signal [34].

Cell based assays for NRs range from reporter gene assays to in vivo recruitment assays. The most reported of these is the GAL4 reporter assay. This assay takes advantage of the fact that the GAL4 response element of yeast does not exist in mammalian systems.

This assay has been thoroughly reviewed [34] and is outlined in Fig. 5. In brief, a cell is transfected with a reporter plasmid consisting of a GAL4 response element upstream from luciferase. NRs are produced as chimeras consisting of the GAL4

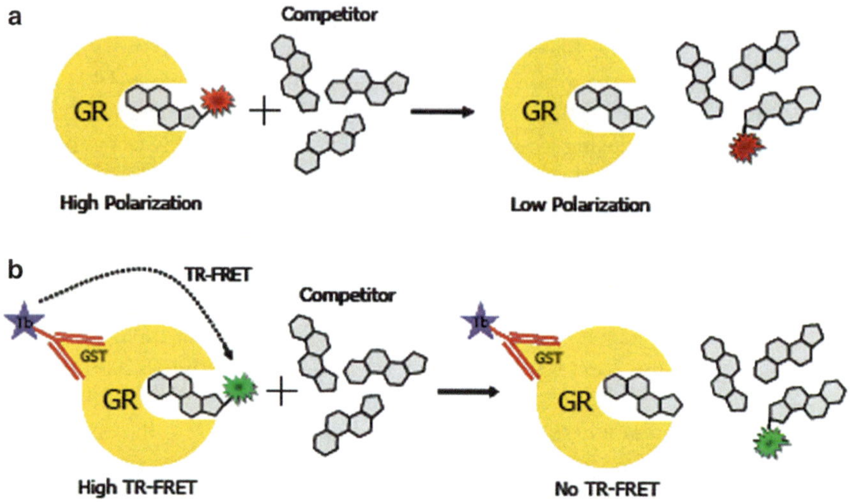

Fig. 4 In vitro nuclear receptor assays. **a** FP assay for compounds compete against a fluorescently labeled steroid for binding to the nuclear receptor. **b** FRET assay for inhibition of coactivator recruitment

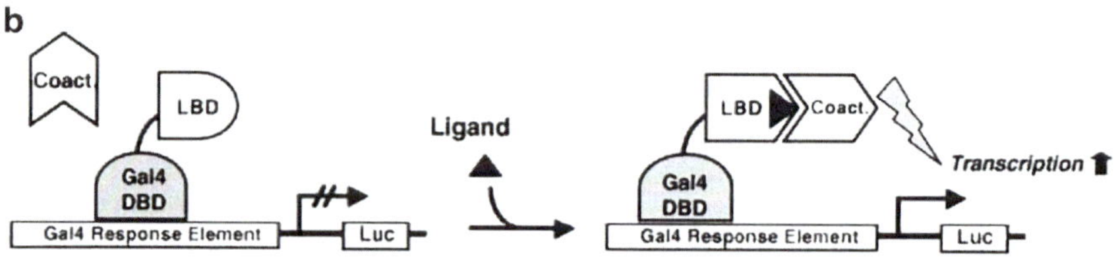

Fig. 5 Gal4 cell-based assays. Cells are transfected with a construct containing the DNA binding domain of Gal4 (*Gal4 DBD*) fused to the ligand binding domain (*LBD*) of a receptor on interest. A second plasmid containing the Gal4 response element upstream of a reporter gene (*LUC*) is cotransfected. Ligand association with the LBD recruits transcriptional coactivators (*Coact*) resulting in increased transcription of the luciferase reporter gene (*LUC*)

DNA binding domain (DBD). When ligand binds to the NR, it translocates into the nucleus, the DBD binds to the GAL4 response element (GAL4RE) driving luciferease transcription. The produced luciferase provides a subsequent detectable fluorescent signal.

3.3.2.4 G Protein Coupled Receptors

GPCRs constitute the largest family of drug targets. Current estimates suggest greater then 80 drugs that have a GPCR as their target, generating over $100 MM in annual sales. With the sequencing of the human genome, there have been over 5,000 GPCRs identified. They account for the majority of best-selling drugs and about 40% of all prescription pharmaceuticals on the market. Notable examples include Eli Lilly's Zyprexa, Schering-Plough's Clarinex, GlaxoSmithKline's Zantac, and Novartis's Zelnorm.

Most HTS labs are performing cell based functional assays for GPCR targets. After ligand binding which stimulates the receptor, intracellular signaling pathways are activated and this leads to a decrease (coupling to a Gi protein) or increase (via Gs) of cyclic AMP (cAMP), or an increase in intracellular Ca, or changes in levels of inositol phosphate (IP). These secondary messengers can easily been monitored in several assay formats.

There are cAMP assays that can be measured using a variety of techniques [35], including HTRF (CisBio) [36], bioluminescence (cAMP-Glo, Promega) [37], TR-FRET (CisBio) [38], LANCE [39], Alphascreen (Perkin Elmer) [40] and Enzyme Fragment Complementation (EFC) [41] (Hithunter, DiscoverX). This review will focus on the LANCE, the TR-FRET and ECF assays as they are the most used in our laboratories.

The LANCE cAMP assay is a competitive assay in which cAMP produced by the cells competes with fluorescent-labeled acceptor cAMP for a cryptate tagged donor antibody. The principal of the assay is shown in Fig. 6. On the left streptavidin conjugated Europium binds to biotinylated cAMP. An antibody labeled with the fluorescent dye Alexa binds to the cAMP, bringing the donor and acceptor into close proximity, and energy transfer occurs. When the cell releases cAMP, it competes with the biotin-labeled cAMP for the antibody, and a signal decrease is observed. In the TR-FRET assay the antibody is directly labeled with either Eu or Tb. In this format an increase in cAMP also causes a decrease in signal.

The third format is the Hithunter assay and is based on the ability of two enzyme fragments to recombine into active enzyme. The principle behind the assay is shown in Fig. 7.

In this assay, free cAMP molecules from the cell compete for antibody binding with a labeled enzyme donor (ED)–cAMP conjugate, which contains a small peptide fragment of β-galactosidase. In the absence of free cAMP, the ED–cAMP conjugates are captured by the cAMP-specific antibody and are unavailable for complementation with the enzyme acceptor (EA), resulting in a low signal. In the presence of free cAMP, antibody sites are occupied, allowing the ED–cAMP conjugate to complement with EA, forming an active β-galactosidase enzyme; substrate hydrolysis by this enzyme produces a chemiluminescent signal. The signal generated is in direct proportion to the amount of free cAMP bound by the antibody. The assay format can be miniaturized to 2μL volumes [42].

A direct comparison of all three assays generated similar statistical data for a variety of receptors (http://las.perkinelmer.com) but the ECF was less sensitive by an order of magnitude. The LANCE was reported to be the platform of choice for assays lasting more then 24 h.

Another popular assay for GPCR activation is to measure the increase in intracellular Ca^{2+} that occurs upon activation. GPCRs on the cell surface produce inositol triphosphate (IP3) via the action of Phospholipase C (PLC). IP3 stimulates calcium channels called IP3 receptors on the endoplasmic reticulum, which raise

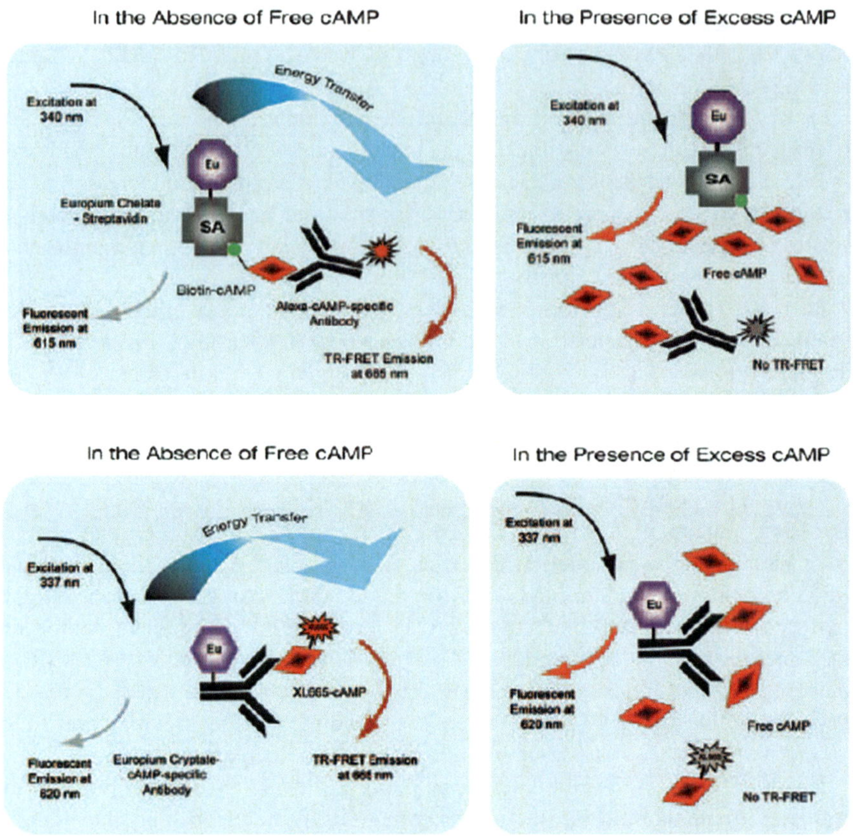

Fig. 6 Assay formats for LANCE (*left*) and TR-FRET (*right*) for the detection of cAMP

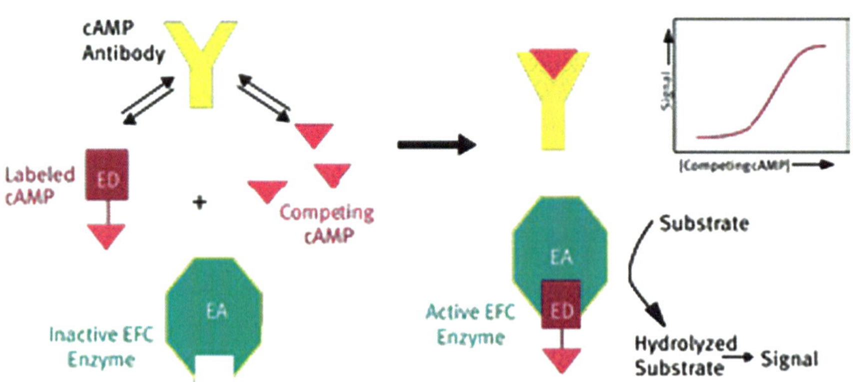

Fig. 7 Hithunter (CisBio) enzyme fragment complementation assay for the detection of cAMP. See text for detailed explanation of the assay components

the concentration of Ca^{2+} ions in the cytoplasm dramatically. GPCR targets that couple via Gq naturally produce an increase in intracellular Ca^{2+} that can be measured using calcium-sensitive dyes and a FLIPR™ instrument. GPCR targets that naturally couple via GI/o can be adapted to respond to agonist with a ligand-dependent increase in intracellular calcium by the use of chimeric G-protein or by the introduction of an over-expressing promiscuous G-protein (Gα15 or Gα16).

These assays are easy to perform. Cells are plated 24h before the assay, preloaded with the dye, and allowed to incubate for a period of time, generally 1 h. Agonist and/or antagonist is added to the cells, the cells are incubated, and then read on the FLIPR.

The assay format works because the dyes are sensitive to the concentration of Ca. In 1985 Roger Tsien's group at Berkley were looking for molecules whose excitation spectrum changes in the presence or absence of Ca^{2+} ions. They discovered a molecule called fura-2, and regardless of whether fura-2 binds Ca^{2+}, it emits light at ~510 nm. However, the wavelength at which it absorbs light is dependent on whether Ca^{2+} is bound. In the absence of Ca^{2+}, fura-2 is excited by 360 nm light; when saturated with Ca^{2+} ions, fura-2 is excited by 330 nm light. Therefore, if you compare the intensity of 510 nm light that is emitted when you shine 360 nm light on your biological sample to the intensity of 510 nm light that is emitted when you shine 330 nm light on your sample, you can calculate the concentration of Ca^{2+} ions. Ca-sensitive dyes have evolved from the original fura-2 to be brighter, better able to penetrate cells, and more stable. The source of dye which works best for a particular assay is most often empirically determined.

One additional technique for measuring GPCR activation is the measurement of IP1 [43]. GPCR Gq stimulation is known to induce PLC activation and trigger the IP cascade. Several metabolites in this pathway, including IP3, have extremely short half lives, making them difficult to accurately quantify. IP1, a downstream metabolite of IP3, accumulates in cells following Gq receptor activation and is stable in the presence of LiCl making it an ideal read out of receptor activation. The IP1 assay from Cisbio is based on competition between released IP1 and fluorescently tagged IP1 for a cryptate-labeled antibody.

Additional methods for measuring GPCR activation are reviewed in Sect. 6.3.3 on High Content Screening.

3.3.2.5 Ion Channels

Ion channels exist to control the concentration of ions inside and outside the cell. They are specialized channels, usually specific for a particular ion. A famous example is the hERG channel, which is responsible for cardiac output. Ion channels exist as either voltage gated or ligand gated, defined by the stimulus that gives rise to the change in potential. Most assays for ion channel activity involve the measurement of changes in membrane potential or ion flux by using a dye that is sensitive to changes in pH or ion concentration [44].

Fluorescence-based methods do not directly measure ionic current but, rather, measure either membrane-potential-dependent or ion-concentration-dependent changes of fluorescence signals (from fluorescent dyes loaded into the cytosol or cell membrane) as a result of ionic flux. Because fluorescence-based methods give robust and homogeneous cell population measurement, these assays are relatively easy to set up and achieve high throughput.

Fluorescent voltage-sensor dyes are used to measure voltage changes across the cellular membrane through either the potential-dependent accumulation and redistribution [45, 46] or the FRET mechanism [47]. The lipophilic, negatively charged oxonol dyes, such as *bis*-(1,3-dibutylbarbituric acid) trimethine oxonol [$DiBAC_4(3)$] are examples of redistribution dyes. DiBAC-based voltage-sensitive dyes have low fluorescence in an aqueous environment, but show increased quantum yield upon binding to hydrophobic intracellular molecules.

In FRET-based voltage sensors, different negatively charged, membrane-soluble oxonol dyes are used as voltage-sensing FRET acceptors – *bis*-(1,3-dialkyl-2-thiobarbituric acid) trimethine oxonol [$DiSBAC_n(3)$]. The FRET donors are coumarin-tagged phospholipids that are integrated into the outer leaflet of the membrane when loaded into the cells. An increase or decrease of FRET in response to membrane hyperpolarization or depolarization produces fast, ratiometric changes. The ratiometric nature of the assay helps to eliminate many of the artifacts associated with DiBAC assays. Unlike DiBAC assays, the use of phospholipid-anchored FRET donor restricts the location of FRET in the plasma membrane, which ensures the measurement of potential changes that occur at the cell membrane, rather than in other subcellular compartments such as the mitochondria.

3.3.2.6 Protein–Protein Interactions

As targets become increasing more complex, the assays used to measure them become complex as well. This is most evident in the current assays which measure protein–protein interactions. Most involve tagging the interacting partners with a variety of either FRET partners or F–Q pairs, or measuring via FP changes [48].

Several more interesting assays involve tagging one of the proteins with either biotin or GST, binding it to a bead, and monitoring the FP signal generated from aggregation of a fluorescently tagged partner. Another assay technique involves the use of AlphaScreen beads to tag each of the partners with either an acceptor or donor bead. Research into neurodegenerative diseases such as Parkinson's and Alzheimer's Disease employ a variety of these techniques to measure protein aggregation.

3.3.2.7 Phenotypic Assays

Phenotypic screens seek compounds with a specific effect on cell physiology, by adding compounds to living cells, and scoring for an effect. The measurable parameters that can indicate these effects can include changes in morphology, staining intensity, spatial attributes of the cellular nuclei, and microtubule formation. Screening small molecules against a target in vitro does not provide information about the effects of test compounds on cellular functions. Phenotypic screening, although much slower to carry out than biochemical screens, provides information about effects on cell or tissue structure or function and therefore can be used to eliminate at an early stage compounds that are toxic or do not produce the desired cellular response.

With complex phenotypic assays, a main challenge is managing and integrating biological information – such assays are designed to probe networks of interactions that give rise to complex processes [49]. Small molecules are not simply exposed to one target, but rather a broad array of networking targets, increasing the complexity of data interpretation. Large target spaces, such as signaling pathways, metabolic networks, or entire expressed proteomes, demand increasing complex roles for information science in interpreting HTS results.

Despite the complexity of the experiments and the enormous data manipulation necessary, complex biological pathways, as well as new drug targets are being identified by this method. Examples include screens for compounds that arrest cells in mitosis, that block cell migration, and that block the secretory pathway [50], or assays with primary T cells from PLP TCR transgenic mice for their inhibitory activity on the proliferation and secretion of proinflammatory cytokines in PLP-reactive T cells [51], and identification of small-molecule inhibitors of histone acetyltransferase activity [52].

3.4 Screening Sample Management

Despite 20 years of practice the proper content and conservation of screening libraries is still the subject of vigorous debate at conferences and to a lesser extent, in the literature. While there is growing consensus that the physical properties of candidate screening samples should be restricted, the details vary from library to library [53]. While there is growing consensus that cool, dry storage conditions tend to best preserve the value of screening samples, the stringency with which these conditions are maintained differs widely. While there is consensus that automated systems add value, the extent of integration of the automation varies from facility to facility.

3.4.1 Solvent

Screening groups have experimented with several different solvent systems for manipulating and storing compounds. Solubilization of compounds in an organic solvent converts dry powders, oils and gums into liquids with more uniform properties that can be more readily and quantitatively transferred from container to container in massively parallel fashion with automated precision pipettors. Once transferred into the penultimate container, usually a 96- or 384-well sample source plate that will be used as the source of test samples for a matching format 96- or 384-well assay plate, it might be more convenient to store the samples in a dry state until use. This led to early experiments with volatile solvents such as acetone that could be easily removed after the compound solutions were arrayed into plates. The mechanics of mixing in miniature containers can be limiting, however, and concern about the recovery of dry compounds from plates led to the development of complex humectants to aid resolubilization of the dried compound films. Despite reports of success with these formats, almost all screening facilities dissolve their screening libraries in the "pseudouniversal solvent," dimethyl sulfoxide (DMSO) and store the solutions until use. Given that the potency of most HTS hits is low to mid micromolar and that most cell-based assays will not tolerate a final DMSO concentration of much more than 0.3% and that the smallest volume transfers routinely achievable with standard pipettors are around 1μL, most compound collections are dissolved at a stock concentration of 10–30 mM. This is a compromise that exceeds solubility limits in DMSO for many compounds, but is the most dilute stock concentration that can still be successfully transferred to a cell-based assay at 10μM with 0.3% residual DMSO, using conventional pipettors.

3.4.2 Storage Conditions

DMSO is a relatively weak oxidizing agent, but there generally seems to be little concern over the potential for chemical modification of compounds stored in DMSO solution. Rather, the major concern with storage in DMSO solution is the hygroscopic uptake of water and the resulting degradation of compound solubility in "wet" DMSO solutions [54]. In the rush to HTS during the early 1990s, many compound libraries were dissolved and handled under ambient conditions of temperature and humidity that inevitably led to substantial water uptake with significant effects on solubility that were only exacerbated by the frequent uncontrolled freeze/thaw cycles that these solutions endured over succeeding years of multiple reuse. It is scant wonder that LC/MS analysis of samples from several traditional HTS libraries indicated loss or degradation of 20–50% of the samples [55]. However, experiments with traditional libraries are de facto observations of "wet" DMSO solutions. Some studies have shown that "dry" DMSO can better preserve test sets of compounds prepared and observed under controlled conditions and that multiple freeze/thaw cycles have little effect under these conditions [56]. In response to the somewhat sparse literature and unpublished internal studies, several pharmaceutical companies have adopted procedures and automation devices designed to dissolve their collections in "dry" DMSO and to maintain the DMSO solutions in a cold, dry state until use. Given the evidence of the deleterious effects on compound solubility resulting from even low level water uptake, this seems like a prudent precaution, although the cost-effectiveness is difficult to quantify. The choice of a storage temperature below the freezing point of the DMSO solutions would also seem prudent, given that rapid freezing should arrest the crystallization that presumably precedes precipitation. One factor in the current uncertainty about storage conditions is the significant cost associated with maintaining an effective quality control program to monitor the purity and integrity of screening libraries ranging from half a million to a million and a half compounds. Despite reports of automated analytical systems, the time and expense it takes to determine the status of every compound on a rotating schedule is more than most organizations are willing to spend [57–59]. Ultimately, as low volume liquid handling instrumentation has improved and the miniaturization of precision labware has evolved, the preparation of multiple replicate copies of small aliquots of the screening library in plates and microtubes designed for one-time use may obviate some concerns surrounding storage conditions and freeze/thaw cycles.

3.4.3 Plate Format

The choice of storage format for compounds was traditionally driven by the assay format, which was limited to those formats compatible with the liquid handling and plate reading formats supported by the instruments used for HTS. When 96-well plates were the rule for HTS, ample volumes of stock compound solutions could be maintained in deep well 96-well format plates to supply screening samples for multiple campaigns using simple one-step liquid transfer processes. These stock deep well plates were often stored under poorly controlled conditions such as ambient humidity at 4°C, which is above the freezing point for "wet" DMSO solutions. Much of the concern about compound degradation and precipitation arose from observations made on plates like these. As 384-well formats were widely adopted by HTS groups and supported by automated liquid handling workstations capable of reformatting 96-well plates to 384-well plates, the volume limitations of 384 well, deep well storage plates and nascent concerns about the effects of multiple freeze thaw cycles led to the development of individual microtubes that contained over 1 mL of solution and could be arrayed in tube racks that mimicked the standard 96-well plate format. The microtubes could be stored in automated systems under environmental conditions designed to preserve the compounds and they could be accessed individually so that HTS hits could be reformatted for confirmation testing without thawing an entire 384-well plate of samples. Two-dimensional matrix codes on the bottom of each tube enabled each tube and the compound it carried to be tracked through every process. Eventually, picotubes were developed that afforded the same benefits in 384-well format. The use of 1,536-well plates has not been widely accepted for HTS, despite the availability of pipettors, dispensers, and plate readers that are compatible with this format. Although most studies indicate that most assays can be reconfigured for typical 1,536-well volumes, it seems that low volume 384-well plates will also function well at these volumes and afford much of the cost savings associated with miniaturization. A more significant barrier to widespread implementation is the extravagant expense of retooling not only the HTS groups, but the assay development groups that would need access to 1,536-well technology to develop HTS compatible assays and the compound management groups that would need to supply compound source plates or assay ready plates containing compounds in 1,536-well format for HTS as well as continuing to supply 384- and 96-well format plates for the rest of their clients. Most screening sample management groups currently dispense compound solutions in both 96- and 384-well plates and maintain cold, dry compound solution stores in individual two-dimensional bar-coded tubes compatible with 96- and 384-well plate formats.

3.4.4 Logistic Strategy

Screening sample management is obviously on the critical path for all screening operations. It is usually much more efficient to centralize this function, even when HTS is decentralized, and when HTS is a central operation, it is usually most effective to colocate the two functions. This minimizes communication and sample transfer delays. Because the two groups use much of the same liquid handling and information technology (IT) infrastructure, they can help support each other's training and method development, ensuring cooperative behavior. At Wyeth, the main dry compound repository and automation for dissolving compounds in DMSO under dry conditions is located 75 miles away from the central screening sample management and HTS facility. This requires routine shipment of new screening samples dissolved in dry DMSO and stored frozen under dry nitrogen in 1.4-mL microtubes from one facility to another. Once the samples are received and stored frozen in a random access automated freezer system maintained under nitrogen, all subsequent screening sample management operations from creation of screening sample source plates through HTS hit confirmation and provision of samples to client labs for hit to lead activities are performed in this central facility, co-located with the HTS facility.

Every year or two, the screening sample management group creates a new edition of the screening library by individually accessing the more than a million 1.4-mL tubes containing the main stock of each compound solution in DMSO. The outputs of this process are replicate copies of 384-well plates containing small aliquots of high concentration compound stock in DMSO that will be diluted immediately before use to provide samples for screening and replicate copies of 100-μL tubes containing just enough high concentration compound stock in DMSO (about 15 μL) to provide material for confirmation assays for HTS. These plates and tubes are stored frozen and sealed under dry nitrogen. Individual sets of the replicate plates comprising the screening library are sealed under dry nitrogen in stacks of 26 ready to use plates in automation friendly metal canisters and manually stored in a walk in freezer with rolling shelves and a bar-code inventory system. The appropriate canisters are removed from the freezer just in time before use and staged next to the appropriate screening platform on specially fabricated carts. Sample source plates are used and discarded and the empty canisters returned for reuse. The screening sample management group must maintain an adequate stock of single use sample source plates to feed 8 concurrent screening campaigns which each consume from 50 to 300 sample source plates 5 days a week. The sealed tubes are stored in an automated system maintained under cold, low humidity conditions where they can be rapidly individually accessed and reformatted in a variety of plate configurations for follow up dose response testing of HTS hits by the HTS group. Each HTS campaign generates hundreds to thousands of hits that require at least two stages of follow up assays in HTS, each stage of follow up testing requires a new round of tube picking and reformatting. The steady state process required to support 30–50 campaigns a year can be estimated, but in the real world, with 8 concurrent campaigns, several times a year there will be a peak demand to serve multiple campaigns that arrive at the end stage simultaneously. Coping with this variable load while maintaining uniform levels of service for all the various activities requires close coordination of resources and strategic logistic planning, including reliable supplies of consumables and adequate reserve stocks as well as the same level of automation engineering and facilities support that are required for the HTS operations.

3.4.5 Automation Systems

It should probably not be surprising that the first large scale automated systems designed for screening facilities in the early 1990s were not designed to perform biological assays, but rather screening sample management activities. In essence, the explanation is simple; every screening campaign makes the same demands on the screening sample management group. In comparison with the diversity of different assay formats that HTS systems must support, the well-defined mission of screening sample management simplifies the design of automated systems that support it. Given the maturity of this effort, the state of the art in automation for screening sample management has not changed much since the last comprehensive review [60]. The market for large integrated systems that can adequately serve a central HTS facility is still dominated by a few large vendors, but smaller, modular systems that can be expanded to rival the capacity of the large integrated automated systems are also available. The need for speed and accurate inventory control while stamping out replicate HTS sample source plates or while picking and reformatting samples from tubes to plates is still the real driving force for the application of automation, but most vendors have added environmental control to the list of features available on their systems. Any system that maintains samples in a frozen state needs to control and minimize humidity in order to prevent ice deposits from interfering with the automation. From that perspective, maintaining a dry nitrogen environment in the cold box is simply a more effective way of excluding moisture.

The real challenge for vendors of automated systems for screening sample management is not the hardware or automation control software needed to move plates, tubes and liquids through their complex, highly sophisticated storage and liquid handling systems, but rather building an information highway between their own proprietary IT systems and the often highly idiosyncratic and sometimes homegrown IT systems used to register samples and track inventory by their corporate clients. There are no universal conventions for terminology and sample tracking and genealogy, and the business rules that define these conventions vary widely from corporate client to corporate client. Factor in the possibility that the client corporation may still be attempting to reconcile multiple legacy systems and it is easy to see why the automation is often the easier deliverable.

4 Deliverables

4.1 Output

In the simplest of terms, an HTS campaign is a process to evaluate thousands to millions of chemical compounds in an assay to identify and quantify activity against a drugable target. This process delivers a compiled list of the active compounds with the associated data on potency, selectivity, specificity, and functionality. This informational data package is assimilated by project team biologists and chemists to enable their decision process on where to deploy chemistry and biology resources over the course of a multiyear lead optimization program. Yet, this process of generating raw data and transforming it into meaningful information useful to others is anything but simple.

HTS campaigns usually consist of several phases – HTS assay development, adaptation of the HTS assay onto robotic platforms, validation of the automated assay, and then sequential evaluation of a sample collection in different assays. The evaluation assays often consist of the primary screen retests to confirm activity, counter screens to eliminate nonspecific interference with the assay format or reagents, selectivity and potency IC_{50} determinations, and often some cell-based functional assay. Many HTS laboratories now include toxicity determination on the most active samples, cross screening against panels of related members of the target class, plus purity and integrity testing of the chemical samples to ensure correct structure. At each step in the process the activity of the unknown sample is compared to internal standards and the performance of each assay assessed by inclusion of internal QC samples and plates. As discussed in the following section, sophisticated statistics are used at each step to monitor assay reliability during the entire campaign and allow identification of all significant active samples.

This complex data set is analyzed to identify and group samples into categories based on initial activity in the primary assay, confirmed activity indicating a true active, selective compounds based on the IC_{50} determinations, and activity in a cell-based assay. Applying cheminformatics tools, each grouping of compounds can be clustered based on chemical structure suggesting preliminary SAR around a chemotype. This analysis is also useful in identifying structural classes possessing known activity that should be avoided, appears less tractable, or may present intellectual property challenges. The final deliverable is a rich data package most often delivered using sophisticated visualization tools allowing combination of quality control aspects, performance of standards, statistical analysis of the results from each assay in the campaign, and clustering of compounds by potency, selectivity and structural class. Additional value is added by annotating the compounds possessing the desired activity with historical data from previous HTS campaigns against both members of the same target class and members of target classes.

4.2 *Statistics*

A basic understanding and appreciation of the value of statistics is an essential skill for HTS. A recent review written from an engineering perspective provides an excellent introduction to the basic concepts, especially as applied to assay design and quality control, the two areas in screening where the application of statistics is both necessary and relatively straightforward [61]. Another recent review, written from a more academic point of view, invests more effort in an attempt to answer the important but thorny question of how to apply statistics to HTS hit selection, whereas in most HTS facilities, every compound is tested only once, at one concentration, for an "*n*" of 1. The obvious answer is to test every compound in the screening library at least twice, if not thrice, but given the magnitude and expense of testing every compound even once, multiple, replicate tests of the entire library will probably not be considered cost-effective in most HTS facilities [62].

4.2.1 The Zhang Factor

Nearly a decade has passed since the venerable Zhang factor was introduced to screening [63]. It is a straightforward method to calculate an easily interpreted, dimensionless number that reflects both the variability of the high and low signals and the separation between the means of the high and low signals of an assay. In other words, it evaluates both the noise and the bandwidth of an assay. When applied to multiple replicate tests using a well characterized reference standard control compound or biological reagent, the resulting Zhang prime factor can be used as a measure of assay quality. When applied retrospectively to the aggregate results from a large set of screening library samples, the Zhang factor can help describe how well the assay performed with real samples under production conditions. Comparison of the Zhang factor to the Zhang prime factor for a given assay can aid understanding of how the composition of a given screening library yielded results that deviated from the ideal behavior of the reference standard. Because the Zhang prime factor is a dimensionless number, it can be used to assess the relative quality of different assays independently of the quality of the screening libraries to which they are applied. Zhang reported that screening assays with Zhang factors in the range of 0.2–0.6, where a factor of 1.0 represents perfection, were commonly encountered in their HTS operation. At Wyeth, screening assays that have Zhang prime factors of 0.4 and higher usually have good outcomes. The Zhang factor is a versatile tool that can be applied to reference standards at each separate concentration across the full range of dose response. It is important to remember that the Zhang factor was devised as an aid to assay development and prediction of the probability of success in HTS and that, like any tool based on statistics, it derives its power from multiple independent observations that are free of systematic error.

Indeed, one of the most thorough retrospective analyses of HTS data suggested that in most HTS operations, systematic effects far outweigh stochastic probability distribution as the leading cause of uncertainty in hit identification [64]. The use of standard statistical tools, including the Zhang prime factor, to assess assay quality depends on replicate tests and the elimination of systematic error. These conditions are difficult to meet. Careful analysis of HTS data from 20 campaigns that screened libraries of 0.5–1.6 million compounds per campaign showed that systematic artifacts were typically observable in most campaigns even after egregiously unacceptable results were identified and rejected by human inspection of visual representations of the data from each plate. The artifacts in question are known to all screeners and included edge effects commonly caused by uneven heating and/or evaporation of the edge wells on a plate, column and row artifacts that can be attributed to the characteristic dispense patterns of reagent and cell dispensers and/or plate washers when they are slightly out of adjustment, plate location specific artifacts that result from failure of individual pipette tips to deliver reagent to individual wells in successive plates due to damage or blockage during the assay, repetitive patterns of high or low signal in areas of successive plates caused by plate reader bias or even manufacturer defects in labware. While assay developers and HTS groups strive to remove as many root sources of systematic error as possible before beginning an HTS campaign, there is a point of diminishing returns at which an informed decision to proceed despite less than perfect statistical performance will deliver an adequate return on investment and clear the platform for use on the next project. And while most HTS facilities perform extensive preventative maintenance on their HTS platforms between campaigns, there is ample opportunity for slight equipment malfunctions to emerge over the course of a campaign.

It might be expected that the use of Zhang prime factors throughout the campaign would provide the first alert of these artifacts, but the application of Zhang prime factors to quality control analysis in many HTS operations is flawed by the fact that, at Wyeth, the only cost effective mechanisms for loading the replicate control wells on every plate that are necessary to calculate individual Zhang prime factors is to place them on the outer edges of the plates, using different liquid dispensers than are used to dispense the screening library samples. The two sets of dispensers have different intrinsic precision and accuracy and, because they operate independently during the assay, the behavior of the reference controls dispensed to the edge wells cannot truly represent the behavior of the samples that are separately dispensed to the interior wells of the plate. One remedy to this situation that is standard practice at Wyeth is to supplement the edge well controls on each plate with separate quality control plates that are interspersed throughout the batch run and that carry multiple replicate wells dispersed over the entire plate that represent the high and low controls as well as dose responses to reference standard compounds. Because these controls are applied to the plates with the same pipettors that are used to dispense the screening library samples, they usually present a more accurate representation of the quality of the sample results, but they only give a snapshot of assay performance at intervals throughout the batch run and inspection of the results of each plate individually is still necessary to catch rogue artifacts so that those plates can be retested. Needless to say, this obsessive quality control requires robust IT support.

Another major limitation in the use of the Zhang prime factor is that, for many targets of interest in HTS, there are no available small molecule control compounds that have solubility and stability properties suitable for use as reference standards for HTS. In some cases, the biological reagents or control compounds used for assay development are unsuitable for HTS. For some novel targets, there are no known ligands. This often leads to the use of artificial assay conditions to generate a high or low signal during HTS, such as absence of enzyme in an enzyme assay, or the use of a nonspecific agonist in a cell-based assay. While practical, the variability associated with these signals may be a poor model for the variability of signals generated by the biology of interest.

4.2.2 Statistics and Hit Identification

The Zhang factor describes population behavior and assesses assay quality as long as systematic artifacts are eliminated, but it does not identify hits. The use of statistical tools to identify hits based on a single test of each compound requires a few additional assumptions. The key assumption is that most compounds in a screening library will be inactive in most assays, so that the distribution pattern of biological activities observed from the entire population of compounds in the screening library will resemble a normal or Gaussian distribution, the bell-shaped distribution of test results that would arise simply by chance from a sufficient number of control tests in the absence of test compounds, with very few compounds showing significant activity more than three standard deviations from the mean. This assumption that the intrinsic variability in the measurements of the activities of all the samples in the library is essentially the same justifies the use of standard statistical calculations of variability to define outliers showing activity in excess of three standard deviations as hits, requiring further testing. Because, in fact, HTS also assumes that there will be some compounds with real biological activity in the screening library, the first assumption of random distribution is modified by the use of the IQR SD rather than the population standard deviation. The use of the IQR SD proceeds from the assumption that, even though there are likely to be some real outliers at the tail ends of the distribution, the behavior of samples that show activity values within the middle 50% of the distribution (on either side of the mean) will more closely conform to a normal or Gaussian distribution and they can be used to calculate an IQR SD that will more correctly identify outliers lying beyond the three IQR SD threshold as hits. One reason for testing multiple replicates of the LOPAC library during HTS validation is to model the population distribution of activity of a sample set of compounds that is artificially enriched with bioactive molecules. Many of these compounds are drugs at nanomolar concentrations, but show nonspecific activity at micromolar concentrations.

As practiced at Wyeth, this statistics-based mechanism for hit selection results in hit selection thresholds that vary considerably from assay to assay and generate correspondingly varying hit rates of 0.5% to over 5%, averaging around 2%. Hit rates in excess of 5% only occur when the shape of the population activity distribution challenges the operating assumption of normality, and typically require some modification of standard operating procedure because they also challenge the hit picking and reformatting capacity and are usually too expensive to support. As information about the confirmed nonspecific activities and/or relevant cross-activities of each compound in the screening library accumulates in the database, excess hits can often be eliminated as nonspecific for a particular target without further testing. On occasion, an extra confirmation step is added to the process wherein all hits are retested in triplicate at the original screening concentration to eliminate those that fail to confirm their activity before proceeding with the normal dose titration confirmation/interference screen protocol with a reduced number of confirmed actives.

In general, the attempt to retest every compound that passes a statistically defined threshold of activity for each assay and to implement a concomitant assay interference test has been rewarded by recovery of a full spectrum of biological activities and diverse chemotypes in the confirmed hit set. In many cases, the compounds that the medicinal chemists ultimately judge to be the best starting points for lead development exhibited only modest activity (e.g., IC_{50} values of 0.5–5 μM) in HTS.

There are always some assays submitted for HTS that defy attempts to achieve statistical validity. Currently, target validation is usually tasked to specially trained and equipped groups separate from the HTS groups, but the widespread use of RNA interference (RNAi) assays has led to adoption of many HTS and high content screening (HCS) techniques for target validation. The extreme variability of cellular response observed in RNAi assays from well to well and plate to plate in the same batch run would confound standard HTS hit selection techniques, but new data analysis tools have been developed for this specialty application that enable decisions for hit selection, despite the high degree of noise in the results [65]. The technique is called Strictly Standardized Mean Difference and requires multiple replicate samples and multiple replicate controls on each plate to enable sophisticated statistical distinction between subtle hits and noisy negative controls, but offers an alternative statistically based hit selection tool that seems to work even with assays that would fail in standard HTS practice.

4.2.3 Hit Selection by Other Means

Traditionally, HTS groups first determined their resource limited capacity for confirmation assays of HTS hits and then used either high potency or chemical structure considerations to pick that number of compounds for retest. The correlation of activity seen in primary HTS with activity seen in confirmation tests is often poor and the most potent compounds from the primary HTS frequently either failed to confirm their activity or proved to be intractable as starting points for lead development. Hit selection by chemical structure was risky in the days when few library compounds underwent purity and integrity testing before they were selected as hits and often overlooked novel singleton structures in favor of well known clusters with limited potential as intellectual property. As compound management and screening infrastructures increased in capacity, it became more common to set an arbitrary activity threshold and to attempt to retest all compounds that passed that threshold. Arbitrary activity thresholds are sometimes stringently set based on the potency of existing chemical equity, but the desire to identify hits that can compete with existing leads should be tempered by realistic expectations and an appreciation of the opportunity to discover alternate chemotypes with the potential to produce novel leads. In practice, arbitrary thresholds are often adjusted on a project by project basis after consideration of the priority of the target, assay hit rate, capacity for retest and results of whatever in silico analyses are available.

5 Has HTS Been Successful?

5.1 A Pessimistic View

The answer depends on who you ask. The Wall Street Journal, quoting leading scientists, called HTS part of "an expensive fiasco" [66]. They suggest that high throughput sciences, such as combinatorial chemistry and HTS, have slowed drug discovery. They point to the decrease in the number of new chemical entities that have been approved since the advent of these new technologies. One author even compared HTS to "that room full of monkeys pounding on typewriters and eventually producing Shakespeare's Hamlet." On the other side of the debate, most pharmaceuticals companies claim to have discovered several clinical candidates that began from hits in a HTS campaign. Bristol-Myers Squibb, Princeton, N.J., can point to one drug that they have on the market that emerged from an HTS program. Sprycel (dasatinib), an oral inhibitor of multiple tyrosine kinases, was approved for the treatment of adults in all phases of chronic myeloid leukemia with resistance or intolerance to prior therapy, including Gleevec (imatinib mesylate).

The truth most likely lies somewhere in between. Bender [67] published the most quantitative study to date on the success of HTS at Novartis. Several conclusions could be drawn. Particular target types and assay technologies have a great impact on screening success, and this was not always correlated to the number of identifying hits in the HTS runs. For assay formats used a minimum of five times, LC/MS readouts succeed 83% of the time, followed by FP assays, which succeed in 72% of the cases. TR-FRET showed a success rate of 70%, with FLIPR assays (61%), fluorescence intensity readouts (59%), and AlphaScreen (60%) performing averagely, compared against a mean of 57% of all HTS campaigns succeeding. Less successful methods are FlashPlate/SPA (50% success rate) and, in particular, receptor binding assays (25% success rate). Thus, a significant difference between assay types arises. Also, significant mutual dependencies of screening success and target type/assay technology can be observed; while one assay format works well with one target type, this might be completely the opposite for a combination of the same readout technology with a different target type.

Perhaps success is based on picking the right assay format. Kashem [68] compared an HTS for a kinase using three different formats. They observed that 57% of the compounds appeared as positives in all three assays. One can only wonder about the 43% that got away.

5.2 An Optimistic View

The answer depends on how you define success. HTS does not identify "drugs" but identifies the starting chemical structure that enables lead optimization through SAR to develop a lead compound. Thus, critics of HTS are accurate in claiming that it has been unsuccessful in finding "drugs." But the truth is that HTS has been very successful in providing the starting points for 50% or more of the small molecule programs currently in the preclinical pipelines of most of the pharmaceutical companies. Additionally, the number of clinical programs that started years before based on HTS continues to grow.

Recognizing that HTS as a true discipline based on the convergent development of laboratory automation, computational power, larger chemical libraries, and disease target knowledge has been practiced in its current form since the early 1990s, it might be unfair to use marketed drugs as the success metric. Yet, considering that the drug discovery and development cycle takes 12–15 years prior to product launch, HTS has made a significant impact. In a study published by Fox [69] in 2005, the directors of 26 HTS laboratories reported that 104 clinical candidates or marketed drugs were based on original HTS campaigns. This number had increased from 74 in 2004 which leads one to extrapolate that there would be a higher number if a similar survey was undertaken today. Clearly, even using the metric of clinical candidates or marketed drugs, HTS would be considered a success.

Taking a more realistic perspective, the real measure of success is determined by the transition of HTS hits into exploratory (hit-to-lead) chemistry, lead optimization and late stage preclinical activities. The frequency of this progression reflects on the quality of the entire HTS process and the resulting data around active chemical compounds. Many of the well-established pharmaceutical company HTS laboratories have discussed their successes. Burns [70] described the impact of Abbott's HTS over the last decade tracking specific examples of compounds moving through the discovery and development process. He describes the use of NMR, affinity, fragment-based, and more traditional approaches in HTS campaigns leading to several clinical candidates. Bender et al. [67] presented the experience of Novartis' HTS group from the perspective of which target classes, assay technologies, and sample library formats generated the highest success with "success" defined as "project progression beyond the HTS campaign to significant downstream activities." Their data indicates numerous successful campaigns. In their historical review of the evolution of HTS at Pfizer, Pereira and Williams [2] provided several examples of early successes leading to clinical candidates and marketed drugs. These citations plus personal communications from HTS laboratory directors make it clear that in those organizations that have invested in developing infrastructure, processes and technical staff to properly conduct and analyze the output of HTS campaigns, that success is easily measured – and obvious.

An additional success metric often overlooked is the impact of HTS on technology development and the transition of enabling "HTS technology" to upstream or downstream discovery and development processes. The interaction of HTS practitioners with the developers and manufactures of instruments, labware, and reagents over the last 20 years has driven rapid technological advances in all three categories. The practical insight and budgets to support early development coupled to an appetite for cutting edge technology has benefited both the users and the providers. Not only have these technologies had a direct impact on the operational efficiency of HTS groups but laboratories involved in target discovery and validation, disease biology and pharmacology, ADME, toxicology, pharmaceutical profiling, and analytical chemistry have all shared in the benefit of these automated screening technologies and sophisticated data management tools "funded" initially by those focused on HTS.

6 Impact, Challenges, and Future Directions

6.1 Data Management

By the mid 1990s, the several screening groups that eventually coalesced into the Wyeth Screening Sciences department had been running automated workstation screens for about 5 years and the central IT group had provided a homegrown HTS data management application mounted on the corporate VAX cluster that included well-designed, easy to use GUIs in the screening laboratories that enabled online protocol management, barcode tracking, real-time plate-based data reduction and data visualizations with streaming statistics, all loaded onto an ORACLE database with a custom query interface that enabled users to generate SQL statements without understanding SQL and that included some sophisticated cross-assay comparative analysis and report tools similar to those used by the National Cancer Institute at that time. This system was maintained for approximately 3 years and abandoned when key personnel left the company and priorities in the IT group changed and commercial software appeared that promised to perform the same functions with less reliance on local IT support for maintenance and upgrades. The in-house system was not perfect. Introduction of new assay procedures usually required expert IT assistance and insufficient data buffering at the plate readers often resulted in lost data during the then frequent network failures, but given the slow pace of screening at that time, it was adequate. It took over a year of painful effort and close support from the Wyeth IT group before the commercial HTS data management application that was purchased to replace the adequate homegrown system was able to function with reduced capacity. Fortunately, over the past 15 years, many sophisticated options have been introduced to the marketplace.

6.1.1 User Requirements

One reason why HTS data management options are seldom judged to be more than adequate by their users is because these users have continually demanded more functionality and higher peak capacities while retaining creative flexibility and ease of use. The range of activities necessary to develop assays, prosecute HTS campaigns, analyze, communicate and curate the results is much the same as it always was, but expectations have risen such that all the features enabled by relational databases should be available for every conceivable HTS support function. For example, all assay protocols should be securely shared with version control and be fully searchable. Not only must the results from every assay plate be correctly matched to every sample source plate, but quality control data generated from reference samples on each assay plate as well as from QC assay plates interspersed between the library plates must be captured, analyzed, and readily associated with the appropriate sequence of assay plates from that batch process. Ideally, the user should also be able to merge the assay and QC results with meta-data about the batch process including details from the log files of the automation platform that track instrument maintenance and QC records and deviations from the expected schedule to help explain anomalies in the assay results. The conversion of data to calculated results needs to be fully automated to keep pace with the machinery, but provision must also be made for interactive manipulations such as curve fitting, as HTS is increasingly applied to complex kinetic and dose response data. These functions are sometimes available in stand-alone laboratory information management systems that are built and maintained by a team of IT professionals who are dedicated to the support of an individual screening facility, but these are often highly customized one-off creations and are usually isolated from the remaining corporate enterprise network.

All data must, of course, be searchable, retrievable, and subject to acceptance or rejection based on both automated and retrospective QC analysis. Reports must be flexible in content yet easily assembled, intuitively understandable, and easily shared with appropriate security and yes, it would be nice if the reports themselves could become, like the assay protocols, version controlled, searchable documents residing in an appropriately indexed relational database. These are just the basic requirements.

The ultimate goal of all scientists is to analyze their data thoroughly until they are sure that it is valid and to then analyze it in a more global context and discuss it with their colleagues. This workflow requires enterprise level IT tools that can effectively compare and correlate multiple HTS campaigns that generated millions of results from hundreds of thousands of compounds, recognize and chart trends and hierarchies of association and help the scientist visualize them, annotate them, and render the visualizations in media that can be used to share that vision with other members of the team.

And yes, all the other functions that support HTS need access to much the same types of tools for data entry, review, validation, query, retrieval, and report. These include assay development groups, compound registration and sample management groups, automated cell culture groups, engineering support groups, logistics and procurement support groups, and especially those responsible for project management. Obviously no single software application currently serves all of these different data types, but all of these data types need to be managed and they all serve HTS.

6.1.2 Data Management Options

Several commercially available software applications are available that provide the basic functionality for HTS [71]. IDBS offers ActivityBase (www.idbs.com), Symyx (formerly MDL) offers Assay Explorer (www.mdli.com), CambridgeSoft offers BioAssay Enterprise (www.cambridgesoft.com), ChemInnovation Software Inc offer CBIS (www.cheminnovation.com), and Accelrys Inc. offers a collection of data processing and integration modules built around SciTegic Pipline Pilot for plate based data (www.accelrys.com). Genedata has entered this field relatively recently with Genedata Screener (www.genedata.com). In common with many other large pharmaceutical companies, Wyeth uses Activity Base to manage HTS assays and to capture and retrieve HTS data, but supplements Activity Base reporting and visualization tools with customized Spotfire modules (www.tibco.com). Wyeth still largely relies on applications written and maintained in-house for the compound management and inventory functions that serve and protect Wyeth's most precious asset, the corporate compound library.

Many of these products are suites or components of suites that attempt to serve the broader needs of drug discovery, including modules for compound registration and inventory control as well as specialized modules for document management and in silico analyses. A common initiative among large pharmaceutical companies is the development of truly global integrated data repositories that assemble all data types into one vast data-mart with an extremely versatile front end query and reporting tool that can also apply powerful analytic tools to the data and that includes tools for facilitating communication and collaboration. These projects are vast, costly, and extended, but by integrating commercially available modules wherever possible they can be completed and they appear to offer the best chance to realize finally the potential benefits of comprehensive IT support for discovery research within a sustainable IT infrastructure.

6.2 Staff Development

In the late 1980s and early 1990s as HTS evolved throughout the pharmaceutical and biotechnology industries, early practitioners came from diverse backgrounds in pharmacology, natural product discovery, and even medicinal chemistry. They were drawn to HTS for many reasons ranging from a desire to improve the drug identification process to the love of advanced technology development. These adventuresome individuals usually performed all the functions required to complete an HTS from assay development, primary and selectivity screening, sample management, data capture and analysis, culminating in information exchange with a project team. This was achievable when sample collections numbered in the hundreds of thousands and a scientist with a rudimentary workstation could process thousands per week. This increased ability in compound evaluation and data output far surpassed the expectations of most organizations and it was acceptable for an HTS to take up to a year to complete. But as the impact of HTS on the early drug discovery process became evident, the size of sample collections increased, and the technologies related to HTS advanced, organizations raised the expectations of HTS groups and began to dramatically change the role of those scientists involved.

Instead of generalists capable of understanding the minimum basics of all aspects of HTS it became necessary to bring the expertise of specialists to the process to enhance capabilities, to understand rapidly emerging technologies, and to focus on specific phases of the process for efficiency and quality. Pharmacologists, biochemists, cell biologists, molecular biologists, and geneticists became necessary for assay development and proper screen performance. Automation experts with backgrounds in engineering, robotics, fluidics and material sciences were needed to understand and maintain increasingly complex workstations and integrated systems to address issues of surface interactions, labware incompatibility, and subcomponent functionality. Chemists with knowledge of cheminformatics, computational techniques, sample library generation, and proper sample storage and QC techniques joined HTS groups as sample libraries near a million became a common target in many companies. With the increase in both the number of HTS campaigns being performed and in the size of sample collections evaluated, information technologies became a critical expertise for every HTS group to provide effective data capture, storage, and analysis. IT experts to implement and maintain sophisticated computer systems and data bases, biostatisticians to support the complicated analysis of data sets, and bioinformatics specialists capable of data mining to generate knowledge became integral members of HTS groups.

HTS matured from an art to a science and with this evolution the need to attract, train, and retain a diverse scientific staff became a critical success factor for leaders of every HTS organization. The modern HTS organizations present a unique challenge to managers and supervisors attempting to provide career growth and recognition to their staffs while achieving the goal of developing, adapting, performing, and analyzing HTS campaigns in an environment demanding increased output to enable quality decisions which projects advance but with static or decreasing investment of personnel or funding. Several key challenges such as recognition and training are important to overcome in HTS staff development.

Many HTS groups exist either as fully integrated departments or as separate groups within a therapeutic area. This autonomy, though enhancing performance and output, often encourages the perception that the HTS group is a "service provider" not a "collaborator." As the chasm widens, therapeutic area scientists who focus on advancing the understanding of a disease and moving projects through the discovery process towards the clinic can often think of their HTS counterparts as less skilled, less knowledgeable, and of lower importance in the organizational structure. This perception leads to diminished opportunities for authorship on publications or podium presentations at scientific meetings when the HTS contributions to the scientific results are considered routine work from a service group. To overcome this, leaders of HTS groups must foster collaboration across the discovery organization and promote the efforts of their staff as worthy of scientific recognition. Several journals and societies have evolved that are focused on assay development, screening technologies, automation, and data management, offering HTS practitioners an outlet for their scientific endeavors. Though it is often difficult to allocate time to prepare and submit articles or presentations, members of HTS groups need encouragement and support to do so in building their own reputation and career, the respect of the group, and advancing their science.

Equally challenging is providing training for a multidisciplinary team existing as an autonomous entity. Continued training for each HTS scientist in their specialty is required to remain up to date on scientific and technology advances plus refresher courses on core expertise. In addition, a unique challenge in HTS groups is providing the opportunity to cross train in the other key disciplines involved in the total process which not only provides development for each individual but also builds awareness of the complexities and challenges that coworkers encounter and solve during an HTS campaign. Training in each specialty can usually be accomplished by encouraging each scientist to maintain roots in their core scientific discipline within the broader organization and through membership in scientific organizations. This networking and exposure to training opportunities in their discipline focused on application to the HTS process extends capabilities in the HTS group while building the individual's career. The bigger challenge is developing and maintaining a program of cross training within the HTS to build awareness, provide redundancy and offer career development activities. The uniqueness of each discipline as applied to HTS in an organization places the responsibility of establishing and maintaining this ongoing training on the leaders. Job training conducted by staff with knowledge of one discipline to other members of the group, rotation of staff from function to function, seminar series, and mentorship are methods often employed. Again, the challenge is conducting this intensive ongoing training in a pressurized environment focused on short term deliverables versus long term career and staff development.

6.3 New Technologies

The technologies described in this section are new only in the sense that some of them are not yet in widespread use in HTS. In fact, the enormous cost and resources at risk in any HTS campaign and the very significant capital expense required to modify HTS platforms encourage a certain conservatism in many screeners that is completely at odds with their usual fascination with new technology. Fortunately, the users and vendors of HTS equipment and reagents have consistently demonstrated a great capacity for cooperative behavior, forming user groups, online forums, and technology consortia, as well as participating in and publishing the results from advance placement beta programs. These activities have helped to move useful technologies that satisfy unmet needs into the marketplace at an accelerated rate.

6.3.1 Miniaturization and Fluidics

Several pioneering HTS groups have demonstrated that most common luminometric and fluorometric assays can be adapted to 1,536-well volumes of 4–8 μL per well using pin tools and the latest generation of high performance plate readers [72, 73]. Nevertheless, most HTS groups still use predominantly 384-well formats and seem to be moving to 384-well low volume formats in lieu of adopting 1,536-well formats. This conservatism may be due as much to the anticipated cost in time and money to upgrade to new equipment and revise SOPs as it is to the ripple effect that such a change in HTS practice would require of assay development and compound management groups, together with the possibility that because they often serve clients besides HTS, support groups would not be able to make a clean break to 1,536, but would be required to maintain 3 supply streams, 96, 384, and 1,536, rather than the 2 streams they currently must provide. This hesitation to embrace the novelty of 1,536-well formats was already obvious several years ago [74]. Although most 1,536-well HTS is currently accomplished using pin tools, vendors have developed a variety of other dispensers that seem to be adequate for low volume 384- and 1,536-well format assays, including capillary pipettors and acoustic dispensers [75, 76]. Acoustic dispensers can enable picoliter dispensing, but current devices differ from pin tools in that they transfer fluid from only one well at a time. They can also serve as analytical devices to monitor volume and water uptake in plate-based compound collections dissolved in DMSO [77].

Despite much excitement in the 1990s over the potential for microfluidics and nanoengineering to revolutionize HTS, there have been few commercially viable systems that embody the concept. Most notable is the Caliper capillary electrophoresis device (www.caliperls.com), which has proven over many years to be an extremely effective medium throughput screening platform [78]. Despite the excellent quality of the data generated on these instruments, they tend to be more useful as detection devices than in their original mission as laboratories on a chip. There is a common difficulty in realizing the original intention of microfluidic devices to complete an entire assay and detect the result in a single cycle of operation on a chip. Unless the chips are far more massively parallel than the Caliper devices, the biological event of interest and detection of the outcome must occur very quickly, so that the dwell time for each sample on the chip is quite short. This may force the use of reaction conditions that vary significantly from physiological norms and/or standard enzymology practice. The Nanostream microparallel liquid chromatography device processed a 384-well plate in 24-well sections and is no longer on the market [79]. The SpinX device was recently introduced and may offer enough robust, parallel, sample capacity, and programmable complexity to survive (www.spinx-technologies.com). The system is composed of a novel disposable microfluidic cassette that has 32 ports molded into the top edge in the same dimensions as a double row of 16 wells from a standard 384-well microtiter plate. When 12 cassettes are bundled together into a rack, the tops of the bundled stacks present the same dimensions as a standard 384-well plate, facilitating liquid dispensing of reagents and test samples to the ports. Each cassette has a complicated system of microfluidic passages and cavities leading down from the ports to the bottom edge and when the individual cassettes are spun in a special centrifuge, centripetal forces move the reagents and samples through the labyrinth, mixing substrates and enzymes and detection reagents in careful sequence with exquisite timing. The key innovation is that all the connections between the ports and the many channels and cavities that honeycomb the interior of the cassette are created on the fly by shining a laser integrated into the centrifuge through the side wall of the spinning cassette. Lasers are also used to detect the reactions that occur when reagents and samples are mixed, while the cassette is still spinning. The variety of different paths and branch points and mixings that can be programmed into each cassette is large enough to enable a wide variety of different assays. It will be interesting over the next few years to see if the market is clever enough to find a critical use for this clever assay instrument that puts no exotic demands on the current liquid handling capability of most laboratories, minimizes reagent use and wastage and offers an enormous flexibility of assay design.

6.3.2 Label-Free Screening

When asked what is the most exciting new change to come to the HTS lab, the scientists surveyed picked label-free assays. These systems alleviate the need for fluorescent or optical tagging of molecules. Platforms based on acoustic resonance, electrical impedance, microcantilevers, nanowires, and differential calorimetry are beginning to appear in the literature. The current greatest utility is for post-high-throughput screening hit confirmation and mode-of-action studies. Label-free technologies are allowing novel assay formats for the analysis of previously intractable targets.

The most popular label-free systems involve measuring the change in reflectance of a light source after interaction with the sample. Two examples are Surface Plasmon Resonance (SPR, Biacore) and the EPIC system (GE Healthcare). Both involve immobilizing one of the components on a solid support. The surface is then illuminated with light, and the reflected light will have a direction and wavelength consistent with the size of the immobilized target. If a binding partner is captured to the surface, the reflected light will change both its angle of diffraction and its wavelength. This change is proportional to the change in molecular weight of the complex. This is a powerful technique for measuring the on and off rates of binding partners.

6.3.3 High Content Screening and Short Interfering RNA

One of the most exciting new technologies to come to the HTS lab is High Content Screening (HCS). HCS uses a confocal microplate imaging system and living cells to monitor spatially or temporally resolved cellular processes within a single cell. The technology allows for the evaluation of multiple biochemical and morphological parameters in cellular systems. Through combining the imaging of cells in microtiter plates with powerful image analysis algorithms, one can acquire deeper knowledge on multiple biochemical or morphological pathways at the single-cell level at an early stage in the development new drugs.

Almost any process that involves trafficking from one compartment of a cell to another can be monitored by HCS. Assays in the literature are as innovative as the technique, and include monitoring the movement of β-arrestin in GPCR signaling (DiscoverX Path Hunter, Transflour, Molecular Devices), ER to Golgi membrane trafficking, and using FRET to monitor mitotic spindle orientation (http://stke.sciencemag.org).

One technology that has generated a lot of excitement is RNAi. RNAi is a conserved pathway in which short interfering RNA (siRNA) molecules target a complementary mRNA. The siRNA guides the multiprotein complex RISC (RNA induced silencing complex) to the mRNA, whereupon the mRNA is cleaved and degraded, resulting in loss of protein function [80]. This natural process can be enlisted to enable selective silencing of any target gene of interest in human cells, finally empowering mammalian cell researchers with a simple genetic tool. RNAi has the potential to affect every aspect of target identification and validation, making drug development better, smarter, and faster.

siRNA screening represents a paradigm shift in the HTS lab. Small molecules are not screened for their ability to modulate a specific target. Instead, using this technique, scientists can screen for and validate the targets themselves. Coupling siRNA with HCS will give researchers a powerful tool for understanding how gene silencing can effect a variety of cell lines. The reader is encouraged to consult the appropriate references for more detailed descriptions.

6.3.4 Primary Cells

Cell lines used for HTS are usually derived from immortalized human cell lines (e.g., HEK293) or rodent cell lines (e.g., Chinese hamster ovary), that have been manipulated by recombinant DNA techniques to express the target of interest. HTS campaigns using engineered cells may miss some naturally occurring protein partners needed for signal generation. Cultured primary cells, blood platelets, or neurons all possess characteristics that often represent in vivo physiology better than immortalized cell lines. Assays in these cell types could give a more accurate prediction of the activity of compounds in an in vivo model and in the clinic, and thus could contribute to a decrease in attrition in the whole process of small-molecule drug discovery. Primary cells provide a physiologically relevant secondary system for drug screening.

One of the major drawbacks from using primary cells in HTS is the difficulty in obtaining the number of cells needed for a campaign. New technologies have emerged that facilitate cell-based screening with lower numbers of cells than are normally needed, among them the LabChip™ microfluidics systems manufactured by Caliper (http://www.caliperls.com) and the CellCard™ system developed by Vitra Bioscience (http://www.vitrabio.com). Examples of the use of primary cells in HTS have recently been reported. For a more comprehensive review of primary cells lines for HTS, including ways to immortalize primary cells, the reader is forwarded to [81]. Lowering the cell requirements should help allow primary culture cells to be used for cell based assay in drug discovery.

6.4 *Smarter Approaches to Screening*

6.4.1 Focused Libraries

The majority of HTS sample collections are assembled so as to maximize the chemical diversity of the members of the library. In contrast to diversity-oriented libraries, target-oriented libraries are designed to create libraries that are focused around specific chemotypes, molecular species, or classes of compounds. Target-oriented design results in focused libraries with a limited number of well-defined compounds. For example, scaffold compounds can be used as "seed" elements with various functional groups systemically added to the seed scaffolds to create sets of analogue compounds. Target-oriented design methods use 3D shape, 3D electrostatics, pharmacophore models, molecular descriptors, and other methods to generate focused libraries. In addition, if compounds of known 3D structure bind to active sites, they can also be used as seeds for libraries [82].

When building targeted libraries, a common design method is to take existing drug leads and generate neighbors (analogs) of the leads in chemistry space using combinatorial methods and conformational expansions of the lead compounds.

The resulting compound libraries thus include many analogs of the lead compounds, which can be used in additional screens for novel leads.

Focused libraries have been applied primarily to kinases and GPCRs, and less frequently, ion channels. Screening sublibraries where the members are selected computationally, empirically, or both, to interact with a particular class of targets is an extremely effective and cost-efficient way of identifying bioactive chemical matter. It is the moral equivalent to fishing where you know there are fish. However, focused libraries have been shown to be quite promiscuous, having activity against multiple targets. In addition, the chemical diversity is lower in a focused library. This usually means finding what you already know.

6.4.2 Fragment Based Screening

Fragment based screening (FBS) is a technique in which small molecules (up to 200 Da) are screened against a target at in the hopes of finding compounds that can bind to different parts of a protein. It differs from traditional HTS in several ways. Traditional HTS screens molecules of ~500 Da at low micromolar concentrations. FBS looks at smaller fragments at millimolar concentrations in the hope of finding simpler compounds with limited chemical functionalities. This usually comes at the cost of lower affinities (IC_{50}s in the millimolar range) for a target. Figure 8 demonstrates the how the two types of compounds may bind in a hypothetical target.

In Fig. 8 (left) a small molecule will fill several potential binding sites within a target, though not necessarily with good potency. FBS, depicted in Fig. 8 (right), has a greater chance of binding more efficiently because of the reduced steric requirements of the smaller fragments.

A variety of detection techniques are used to measure the binding of the fragments to the targets. Among the more popular are NMR, ITC and DSC, X-ray crystallography, and SPR. The success rate for finding viable leads has been reported as high as 70–80% of targets screened. Card, et al. [83] used FBS for the detection of specific inhibitors of PDE4. Starting with an 80-μM fragment, the authors were able to decrease potency 4,000-fold.

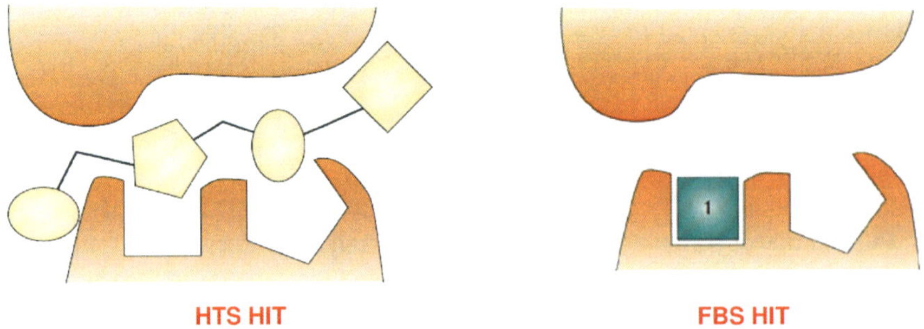

Fig. 8 Schematic representation of the binuing of a small molecule (*left*) or a fragment (*right*) to a hypothetical protein active site

6.4.3 Virtual Screening

Virtual screening involves the screening of millions of compounds in silico to find those with the best chance of binding to the target. Its greatest utility is programs where the structure of the target has been solved by X-ray or NMR techniques. The success of this approach depends heavily on the algorithms used to calculate binding energies. Two major approaches are the quantum mechanical approach and the knowledge based approach.

Quantum mechanical approaches, such as those applied by Merz [84], use traditional quantum mechanics to calculate free energies of binding. Though relatively successful, the computational time required to calculate the energies make this approach impractical to large data sets.

Knowledge based approaches, such as those pioneered by Shakhnovich [85], utilize a more empirical approach. They deconstruct current drug compounds into one- to three-atom segments, and calculate the probabilities that these small pieces exist in these drugs. They then construct new molecules in the binding site of the protein, score them as potential binders, and rank order compounds. The great advantage of this approach is the speed as which compounds can be grown and scored.

Both approaches have their advantages and disadvantages. Each has been successful enough to form the basis of startup companies trying to apply these algorithms to solve complex drug targets. The utility of these approaches will become evident in time.

6.5 HTS in Academics

Over the last 10 years there has been a major initiative to bring HTS into the academic community. The NIH roadmap has a mission to build a better "toolbox" for medical research in the twenty-first century and to empower the research community to use small molecule compounds in their research, whether as tools to perturb genes and pathways, as imaging probes in basic or clinical applications, or as starting points to the development of new therapeutics for human disease.

Part of this initiative is the Molecular Libraries Screenings Centers Network, a collection of screening centers at universities. Additional information on this screening service can be found at http://nihroadmap.nih.gov/molecularlibraries.

A significant number of universities are building HTS facilities. The first was the Institute of Chemistry and Cell Biology, established at Harvard Medical School in 1998. Its purpose was to facilitate the pursuit of Chemical Genetics as an academic discipline. Another interesting example is the screening center at the Neurodegenerative Diseases at Harvard University. It is one of the first examples of a teaching center for drug discovery where not only are screening services provided, but students are taught the sciences of assay development and screening.

What has been the effect of this initiative on the discovery of drugs to fight disease? According to Nature Chemical Biology [86], HTS in industry is focused largely on assaying drugable targets for lead compounds with drug-like properties. In academic research, on the other hand, investigators may be interested in identifying small molecule modulators of biological targets that are not considered drugable or that have no connection to disease. With the broader range of biology under investigation, and without the requirement for optimal pharmacology, it is not necessary, and often not desirable, to limit screening libraries to drug-like molecules. This broad chemical biology purview pushes the boundaries for HTS assays [73] and changes the demands for the composition of chemical libraries [87]. Thus, HTS in academic research is expanding our ability to probe chemical and biological space.

7 Summary

In this chapter we have provided descriptions of the key components of HTS with examples of how this lead identification approach is generally practiced. We have also commented on peripheral issues such as staff development, success metrics, emerging technological trends, plus operational and organizational challenges. Understanding the complexity of the HTS campaign and the multidisciplinary teams necessary in this interactive process is essential for chemists and biologists to harness the potential of this scientific discipline that has evolved over the last 25 years.

Technologies and industries undergo continual change and the next 10 years will no doubt see further change. Advances in technologies, evolution of processes and increases in understanding diseases will enhance the role of HTS in the drug discovery process. We hope that this chapter provides the underpinning to facilitate communications across scientific disciplines involved in HTS to recognize the needed changes and implement necessary solutions for continued success.

Appendix

Accuracy

Closeness of the agreement between the result of a measurement and a true value of the property being measured.

Active

A sample that produces a response or signal above a defined threshold at the tested concentration in a single assay or screen and that has not yet been confirmed by subsequent experiment. Note: when the properties and identity of an ACTIVE sample are confirmed by subsequent experiment it becomes a HIT. Use of ACTIVE and HIT as synonyms is inaccurate.

Activity

The response to a test sample measured in an assay.

Activity distribution

A plot or graphical representation of the number of samples present in each activity range. Often shown as a population bar chart, it is used to provide an overview of the screening results and typically allows the determination of the overall BACKGROUND signal and threshold for selection of ACTIVE samples.

Artifact

An experimental result which is not a manifestation of the phenomenon under investigation, but is brought about erroneously by the particular arrangement of instrument and method.

Assay

(1) An experimentally controlled biochemical or biological system used for the quantitative analysis of perturbations imposed by a test sample; (2) a set of operations having the object of determining the value of a quantity. In analytical chemistry, this term is synonymous with measurement.

Assay control, negative

Experimental conditions designed to produce the minimum signal in an assay. It is typically the signal measured in the absence of a test compound and is relevant to the determination of the BACKGROUND signal. Note: synonymous with "low" control.

Assay control, positive

Experimental conditions designed to produce the maximum signal in an assay. Typically determined with a reference test compound. Note: synonymous with "high" control.

Assay format

Description of an assay in which the plate type (96-well, 384-well, etc.) and assay type (fluorescence, luminescence, etc.) are defined.

Assay validation

Experiments conducted to verify that the output measurement of the assay is reflective of the target activity. Results are compared (where possible) to existing literature parameters such as K_d, K_i, K_m, or EC_{50}.

Automation

Mechanization with process control, where process means a sequence of manipulations. One or several functions in an instrument may be automated.

Background

(1) The amount of a signal produced in an assay or screen in the absence of a test substance; (2) the signal detected from an assay in the absence of TARGET activity; often equivalent to negative control.

Batch

A homogeneous preparation of a reagent/reactant (small molecule, enzyme, CLONE, etc.) produced (synthesized, purified, or otherwise) at one time.

Compound collection/library

A set of chemicals that has been labeled or annotated for easy storage and retrieval and that is available for screening. Consists of compounds synthesized by combinatorial or standard synthetic methods, purchased from commercial or academic sources, or samples of natural products either as pure samples or as mixtures.

Concentration response

Whenever a figure quantifying the affinity or efficacy of a compound in an assay is required, the compound is evaluated at increasing concentrations in order to determine its concentration-dependent effect. The classical thermodynamics usually result in a hyperbolic increase of the assay signal upon linear increase in compound concentration. Thus, compound concentration responses are usually determined using logarithmic serial dilutions, e.g., 10 μM–1 nM (See also K_i, K_D, IC_{50}, EC_{50}).

Counter-screen

A screen in which test samples are assessed against a TARGET for unwanted activity. This target may or may not be structurally or functionally related to the intended target.

Effective concentration 50 (EC_{50})

The concentration of an effector that produces one-half of the maximal response for that system. Usually refers to an agonist in a receptor system, compared to a reference agonist that produces a maximal response in the system.

Efficacy

The extent to which a compound produces a response in an assay, relative to the high and low assay controls. When the compound increases the signal up to 100% or decreases the signal down to 0%, it has a 100% efficacy. When the plateau reaches another intermediary efficacy value, the compound is said to have a partial effect.

False negative

An assay result in which a sample known to be active does not produce either the expected signal or a signal above the activity threshold. FALSE NEGATIVES can occur when an assay lacks appropriate discriminatory power, when the threshold is inappropriately set, or as a result of mistaken identity of the test sample.

False positive

An assay result in which a sample known to be inactive produces a signal or response above the activity threshold. FALSE POSITIVES can occur when an assay lacks appropriate discriminatory power, when the threshold is inappropriately set, as a result of certain physical properties of the substance (e.g., a fluorescent compound in a fluorescence intensity assay, aggregation), or as a result of mistaken identity of the substance.

High content screening (HCS) assay

An assay that produces multiple biological readouts. Most commonly used in relation to the mathematical analysis of an image acquired using an automated microscope whereby analysis algorithms quantify cellular parameters (e.g., number, motility, neurite outgrowth, size, shape) and subcellular events (e.g., receptor internalization, protein translocation, protein expression nuclei shape).

High throughput

A relative term, applied to the generation of a large number of results (e.g., 100,000) in a short timeframe (week or month). Usually achieved by employing a substantial degree of automation.

Hit

A sample that produces confirmed activity above the hit threshold in an assay and whose structural identity has been confirmed. A substance becomes a HIT when the properties of an ACTIVE are confirmed by elimination of FALSE POSITIVE results and ARTIFACTS. Note: in the past, the terms confirmed hit, true hit, and confirmed active were used with this meaning.

Hit rate

The portion of hits that displays confirmed activity in a screen beyond a minimum defined level, the hit threshold. Expressed as a percentage.

Hit threshold

The minimum activity that defines ACTIVES in a PRIMARY SCREEN. It is usually expressed as percentage of inhibition or stimulation. For example, a widely used hit threshold is 50%.

HTS

A method in which a large number of assays (from thousands to millions) are performed and assessed in a relatively short time period. Typically, these assays are carried out in microplates of at least 96 wells using automated or robotic technologies. Note: the rate of at least 100,000 assays per day has been termed "Ultra HTS" (UHTS).

Inactive

A substance that does not produce a response above the hit threshold in an assay at the tested concentration. Note: a substance may also be designated as inactive when attempts to confirm an ACTIVE fail.

Lead

A compound (or compound series) that satisfies predefined minimum criteria for further structure and activity optimization. Typically, a lead will demonstrate appropriate activity, selectivity, tractable SAR, and the potential to be patentable.

Library

(1) A collection of samples (e.g., chemical compounds, natural products, over-expression library of a microbe) available for screening; (2) a set of compounds produced through combinatorial chemistry.

Liquid handler or liquid handling machine

A programmable device that accurately and precisely delivers predefined quantities of liquid to a MICROPLATE. It may be free-standing; incorporated into a WORK-STATION, or part of a fully automated system.

Microplate

Any of a number of plates containing a series of wells in which to store reagents, clones, etc., or perform individual assays. Typically, these plates are constructed of a variety of clear and opaque plastics, and contain 96, 384, or 1,536 individual wells, although 24-well and 3,456-well plates are also available.

Microplate standards

Standards that define the footprint, the height, the flanges and the well positions of 96-, 384-, and 1,536-well microplates.

Module

An individual automated device within a fully automated assay system that usually performs a complete single assay step or procedure. A fully enclosed MODULE may allow for the control of temperature, humidity, and the gaseous environment.

Noise

The random fluctuations occurring in a signal that are inherent in the combination of instrument and method.

Plate format

The number and configuration of wells on a microplate. The most widely used formats are arrays of 96 wells (8×12), or 384 wells (16×24), or 1,536 wells (32×48).

Plate map

The layout of samples and controls configured on a plate during an assay. For example, for a primary screen in 384-well plates, columns 1, 2, 23, and 24 are controls, and columns 3–22 are for individual test compounds, whereas for secondary screening, each row will contain a single compound at varying concentrations.

Precision

The closeness of agreement between independent test results obtained by applying the experimental procedure under stipulated conditions. The smaller the random part of the experimental errors which affect the results, the more precise the procedure. A measure of precision (or imprecision) is the standard deviation.

Primary screen

The initial screen applied to assess the activity of a collection of compounds against a biological target of interest. This SCREEN identifies ACTIVES from a LIBRARY.

Quality control

An operation or series of operations that contributes to the validation of screening results. Such operations include validation of liquid handling devices and plate readers, experiment controls, such as determination of the Z' factor and use of assay controls, and postexperiment controls, such as data analysis validation and database administration. Results of a screen are validated only after a set of quality controls have been performed.

Reproducibility

The closeness of agreement between independent results obtained with the same method on identical test material but under different conditions (different operators, different apparatus, different laboratories, and/or after different intervals of time). The measure of reproducibility is the standard deviation qualified with the term "reproducibility" as reproducibility standard deviation. In some contexts reproducibility may be defined as the value below which the absolute difference between two single test results on identical material obtained under the above conditions may be expected to lie with a specified probability. Note that a complete statement of reproducibility requires specification of the experimental conditions which differ.

Robustness

An assay or screen is said to exhibit ROBUSTNESS when it has a high discriminatory power and produces a low number of FALSE NEGATIVE and FALSE POSITIVE results.

Sample

A portion of material selected from a larger quantity of material. Typically a chemical compound or mixture of compounds submitted to an assay or a screen.

Screen

The execution, analysis, and interpretation of a large number of assays to evaluate the activity of a collection of samples against a target. A screen will often employ automation.

Screen validation

Assay conditions as determined by ASSAY VALIDATION are performed in the chosen PLATE FORMAT with an acceptable signal to background ratio as described by the Z' factor.

Secondary screen

A screen applied to confirm independently actives from the primary screen. A secondary screen may employ an assay that differs in type from the primary screen, e.g., biochemical assay vs cell based assay, or it may be of the same type with different readout.

Selectivity assay

An assay used to determine the relative potency of active or lead compounds towards an alternative target. A selectivity assay (or panel of assays) may include targets of the same family or unrelated targets.

Target

A biological molecule, such as an enzyme or receptor, whose activity and function is the focus of a screen.

Targeted library

Library designed, on the basis of preexisting information, to generate enhanced activity or hit rate against a particular biological target or target class.

References

1. Andersen Consulting (1997) Re-inventing drug discovery: the quest for innovation and productivity. Anderson Consulting, London
2. Pereira DA, Williams JA (2007) Brit J Pharmacol 152:53
3. Glossary of Screening Terms, Society of Biomolecular Sciences. www.sbsonline.com
4. Macarron R (2006) Drug Disc Today 11:277
5. Major J (2002) Curr Drug Dev 7:31
6. Melnick JS, Janes J, Kim S, Chang JY, Sipes DG, Gunderson D, Jarnes L, Matzen JT, Garcia ME, Hood TL, Beigi R, Xia G, Harig RA, Asatryan H, Yan SF, Zhou Y, Gu XJ, Saadat A, Zhou V, King FJ, Shaw CM, Su AI, Downs R, Gray NS, Schultz PG, Warmuth M, Caldwell JS (2006) PNAS 103:3153
7. Soloveva V, LaRocque J, McKillip E (2006) J Assoc Lab Auto 11:145
8. Huang KS, Mark D, Gandenberger FU (2006) Methods Enzymol 414:589
9. Shoichet BK (2006) Drug Disc Today 11:607
10. Leder L, Freuler F, Forstner M, Mayr LM (2007) Curr Opin Drug Disc Devel 10:193
11. Bonetta L (2006) Nature 439:1017

12. Cawkill D, Eaglestone SS (2007) Drug Disc Today 12:820
13. McKillip E, Bracy-Johnson N, Soloveva V, LaRocque J, Morin J, Paslay J (2007) Bioprocess J 2:39
14. Chen J, Lake MR, Sabet RS, Niforatos W, Pratt SD, Cassar SC, Xu J, Gopalakrishnan S, Pereda-Lopez A, Gopalakrishnan M, Holzman TF, Moreland RB, Walter KA, Faltynek CR, Warrior U, Scott VE (2007) J Biomol Screen 12:61
15. Fornwald JA, Lu Q, Wang D, Ames RS (2007) Methods Mol Biol 388:95
16. Perrin F (1929) Photochimie Ann Phys Ser 10:169
17. Weber G (1952) Biochem J 51:145
18. Dandliker WB, Feigen GA (1961) Biochem Biophys Res Commun 5:299
19. Jolley ME (1981) J Anal Toxicol 5:236
20. Jia Y, Quinn CM, Talanian RV (2007) Expert Opin Drug Disc 2:909
21. Kumar R, Clark DS (2006) Curr Opin Chem Bio 10:162
22. Yong J, Quinn CM, Kwak S, Talanian RB (2008) Curr Opin Drug Tech 5:59
23. Zhang W, Shor B, Ker Y (2006) Protein Expr Purif 46:414
24. Huss KL, Blonigen PE, Campbell RM (2007) J Biomol Screen 12:578
25. Vogel KM, Zhong Z, Kun Bi K, Pollok BK (2008) Expert Opin Drug Disc 3:15
26. Von Leoprechting A, Kumpf R, Menzel S, Reulle D, Griebel R, Valler MJ, Büttner FH (2004) J Biomol Screen 9:719
27. Fields B (2001) Meth Mol Biol 151:495
28. Lombard C, Saulnier J, Wallach J (2005) Biochimie 87:265
29. Livingston DC, Brocklehurst JR, Canon JF, Leytus SP, Werly JA, Peltz SW, Peltz GA, Mangel WF (1981) Biochemistry 20:4298
30. George J, Teear ML, Norey CG, Burns DD (2003) J Biomol Screen 8:72
31. Karvinen J, Elomaa A, Makinen M, Hakala H, Mukkala V, Peuralahti J, Hurskainen P, Hovinen J, Hemmila I (2004) Anal Biochem 325:317
32. Levine LM, Michener ML, Toth M, Holwerda BC (1997) Anal Biochem 247:83
33. Lin S, Bock CL, Gardner DB, Webster JC, Favata MF, Trzaskos JM, Oldenburg KR (2002) Anal Biochem 300:15
34. Harrison RK, Lin S, Seidel HM, Zhang J, Zhang L, McGeehan GM (2002) Curr Med Chem Immunol Endocr Metab Agents 2:23
35. Eglan R (2005) Front Drug Des Disc 1:97
36. Ito S, Yoshimoto R, Miyamoto Y, Mitobe Y, Nakamura T, Ishihara A, MacNeill DJ, Kanatani A, Tokita S (2006) Eur J Pharmacol 529:40
37. Kumar M, Hsiao K, Vidugiriene J, Goueli SA (2007) Assay Drug Dev Technol 5:237
38. Persani L, Calebiro C, Bonomi M (2008) Nat Clin Pract Endocrinol Metab 3:180
39. Evans L, Romeo DP (1977) Am J Bot 64:1170
40. Dowless MS, Barbee JL, Borchert KM, Bocchinfuso WP, Houck KA (2005) J Pharmacol 512:9
41. Golla R, Seethala R (2002) J Biomol Screen 6:515
42. Weber M, Muthusubramaniam L, Murray J, Hudak E, Kornienko O, Johnson EN, Strulovici B, Kunapuli P (2004) Assay Drug Dev Tech 5:39
43. Trinquet E, Fink M, Bazin H, Grillet F, Maurin F, Bourrier E, Ansanay H, Leroy C, Michaud A, Durroux T, Maurel D, Malhaire F, Goudet C, Pin JP, Naval M, Hernout O, Chrétien F, Chapleur Y, Mathis G (2006) Anal Biochem 385:126
44. Xu J, Wang X, Ensign B, Li M, Wu L, Guia A, Xu J (2001) Drug Disc Today 6:24
45. Waggoner A (1976) J Membr Biol 27:317
46. Loew LM (1988) Spectroscopic membrane probes. CRC Press, FL
47. Gonzalez J, Tsien R (1997) Chem Biol 4:269
48. Luk KC, Hyde EG, Trojanowski JQ, Lee VM (2007) Biochemistry 46:12522
49. Clemons PA (2004) Curr Opin Chem Biol 8:334
50. Yarrow JC, Feng Y, Perlman ZE, Kirchhausen T, Mitchison TJ (2003) Comb Chem High Throughput Screen 6:279
51. Rossi C, Padmanaban D, Ni J, Yeh LA, Glicksman MA, Waldner H (2007) J Biomol Screen 12:481

52. Hardcastle A, Tomlin P, Norris C, Richards J, Cordwell M, Boxall K, Rowlands M, Jones K, Collins I, McDonald E, Workman P, Aherne W (2007) Mol Cancer Ther 3:1112
53. Shelat AA, Guy RK (2007) Curr Opin Chem Biol 11:244
54. Di L, Kerns EH (2006) Drug Disc Today 11:446
55. Bowes S, Sun D, Kaffashan A, Zeng C, Chuaqui C, Hronowski X, Buko A, Zhang X, Josiah S (2006) J Biomol Screen 11:828
56. Cheng X, Hochlowski J, Tang H, Hepp D, Beckner C, Kantor S, Schmitt R (2003) J Biomol Screen 8:292
57. Popa-Burke IG, Issakova O, Arroway JD, Bernasconi P, Chen M, Coudurier L, Galasinski S, Jadhav AP, Janzen WP, Lagaska D, Liu D, Lewis RS, Mohney RP, Sepetov N, Sparkman DA, Hodge CN (2004) Anal Chem 76:7278
58. Di L, Kerns EH, Chen H, Petusky SL (2006) J Biomol Screen 11:40
59. Schopfer U, Höhn F, Hueber M, Girod M, Engeloch C, Popov M, Muckenschnabel I (2007) J Biomol Screen 12:724
60. Archer JR (2004) Assay Drug Dev Technol 2:675
61. Achyuthan KE, Whitten DG (2007) Comb Chem High Throughput Screen 10:399
62. Malo N, Hanley JA, Cerquozzi S, Pelletier J, Nadon R (2006) Nat Biotechnol 24:167
63. Zhang JH, Chung TDY, Oldenburg KR (1999) J Biomol Screen 4:67
64. Gribbon P, Lyons R, Laflin P, Bradley J, Chambers C, Williams BS, Keighley W, Sewing A (2005) J Biomol Screen 10:99
65. Zhang XD, Ferrer M, Espeseth AS, Marine SD, Stec EM, Crackower MA, Holder DJ, Heyse JF, Strulovici B (2007) J Biomol Screen 12:497
66. Landers P (2004) The Wall Street Journal, CCXLII(37)
67. Bender A, Bojanic D, Davies JW, Crisman TJ, Mikhailov D, Scheiber J, Jenkins JL, Deng Z, Hill WA, Popov M, Jacoby E, Glick M (2008) Curr Opin Drug Disc Dev 11:327
68. Kashem MA, Nelson RM, Yingling JD, Pullen SS, Prokopowicz AS, Jones JW, Wolak JP, Rogers GR, Morelock MM, Snow RJ, Homon CA, Jakes S (2006) J Biomol Screen 1:70
69. Fox S (2005) High throughput screening 2005: new users, more cell-based assays, and a host of new tools. HighTech Business Decisions, Moraga, CA
70. Burns D (2008) J Biomol Screen 13:695
71. Vaschetto M, Weissbrod T, Bodle D, Guner O (2003) Curr Opin Drug Disc Dev 6:377
72. Hodder P, Mull R, Cassaday J, Berry K, Strulovici B (2004) J Biomol Screen 9:417
73. Inglese J, Johnson RL, Simeonov A, Xia M, Zheng W, Austin CP, Auld DS (2007) Nat Chem Biol 3:466
74. Garyantes TK (2002) Drug Disc Today 7:489
75. Taylor PB, Ashman S, Baddeley SM, Bartram SL, Battle CD, Bond BC, Clements YM, Gaul NJ, McAllister WE, Mostacero JA, Ramon F, Wilson JM, Hertzberg RP, Pope AJ, Macarron R (2002) J Biomol Screen 7:554
76. Glaser V (2005) Assay Drug Dev Technol 3:607
77. Ellson R, Stearns R, Mutz M, Brown C, Browning B, Harris D, Qureshi S, Shieh J, Wold D (2005) Comb Chem High Throughput Screen 8:489
78. Wu G, Irvine J, Luft C, Pressley D, Hodge CN, Janzen B (2003) Comb Chem High Throughput Screen 6:303
79. Lemmo AV, Hobbs S, Patel P (2004) Assay Drug Dev Technol 2:389
80. Bartz S, Jackson AL (2005) Science STKE, 2005:39
81. Horrocks C, Halse R, Suzuki R, Shepherd PR (2003) Curr Opin Drug Disc Dev 6:570
82. Miller JL (2006) Curr Top Med Chem 6:19
83. Card GL, Blasdel L, England BP, Zhang C, Suzuki Y, Gillette S, Fong D, Ibrahim PN, Artis DR, Bollag G, Milburn MV, Kim SH, Schlessinger J, Zhang KY (2005) Nat Biotechnol 2:201
84. Gogonea V, Suárez D, van der Vaart A, Merz KM (2001) Curr Opin Struct Biol 2:217
85. Grzybowski BA, Ishchenko AV, Shimada J, Shakhnovich EI (2002) Acc Chem Res 5:261
86. Editorial (2007) Nat Chem Biol 3:433
87. Shelat AA, Guy RK (2007) Nat Chem Biol 3:442

Top Med Chem (2009) 5: 85–124
DOI: 10.1007/7355_2009_3

Published online: 6 Aug 2009

Lead Discovery Using Virtual Screening

Jack Andrew Bikker and Lakshmi S. Narasimhan

Abstract The practice of virtual screening (VS) to identify chemical leads to known or novel targets is becoming a core function of the computational chemist within industry. By employing a range of techniques, when attempting to identify compounds with activity against a biological target, a small focused subset of a larger collection of compounds can be identified and tested, often with results much better than selecting a similar number of compounds at random. We will review the key methods available, their relative success, and provide practical insights into best practices and key gaps. We will also argue that the capability of VS methods has grown to a point where fuller integration with experimental methods, including HTS, could increase the effectiveness of both.

Keywords VS, Virtual Screening, Lead discovery, lead, HTS, Pharmacophore-Based, Structure-Based, Fragment-based, Ligand-based, Docking, Scoring, hybrid workflows, VS strategy, Benchmarking VS

Contents

J.A. Bikker
Department of Chemistry and Screening Sciences, Wyeth Research, 401 North Middletown Road, Pearl River, NY, 10965, USA
e-mail: bikkerj@wyeth.com

L.S. Narasimhan(✉)
Pfizer Global Research and Development, La Jolla Laboratories, La Jolla, CA, 92121, USA
e-mail: Lakshmi.narasimhan@pfizer.com

Abbreviations

ADME Absorption, distribution, metabolism, excretion
FBVS Fragment-based virtual screening
LBVS Ligand-based virtual screening
PHBVS Pharmacophore-based virtual screening
SBVS (Protein) Structure-based virtual screening
SSR Selection to superset ratio
TMVS Text-mining based virtual screening
VS Virtual screening

1 Introduction

Over the last decade, improvements in algorithms for molecular comparison and in docking and scoring, in conjunction with the advent of affordable yet fast computing through clusters of relatively inexpensive processors have made VS a promising strategy to identify novel leads to known and new targets. It is a highly cost-efficient and relatively fast way to leverage limited information on a biological target, namely a small number of compounds that are active against it, or its structure determined to atomic resolution or both, to find additional leads. When successful, this method can often identify leads that are of interest, as defined by the key characteristics of potency, novelty, exploitability, selectivity, and ADME. As a

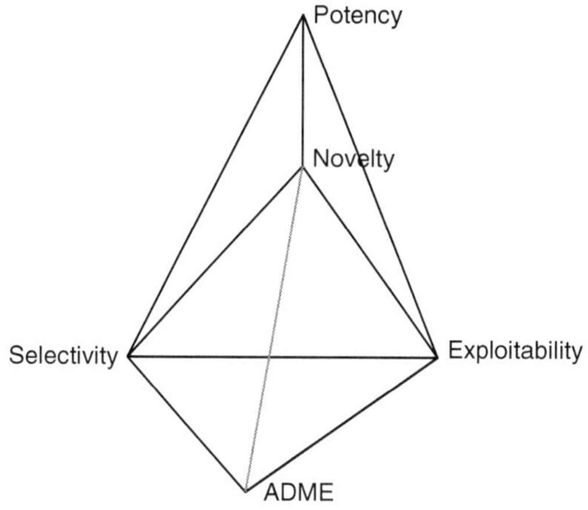

Fig. 1 The desired attributes of a lead molecule. Often, molecules identified by any screening strategy might satisfy optimal criteria for only a subset of these attributes and most laboratories would proceed with a medicinal chemistry campaign banking on improving the rest in a subsequent lead optimization phase

strategy, the software and methods available can be used alone, or in combination with more common high throughput screening (HTS) or fragment screening technologies. As this review will demonstrate, there is ample evidence to show that this technology has been applied to targets in gene families for which inhibitors are known, and to recently identified targets. Success, dependent on a variety of factors and defined in as many ways, is variable, and controlled by what we have termed the zeroth law of screening: *If the compound is not in the screening collection, it can't be found.* However, experience and the literature suggest that, if there are inhibitors of modest potency or better present in the screening collection, a subset may well be found by applied VS methods (Fig. 1).

VS refers to any computational filtering or statistical prediction applied to cherry-pick compounds from a large database. The logical next step is to acquire these compounds for experimental testing. An operational definition of VS, that it is the exercise of ranking molecules by descending order of likelihood of relevant biological activity, regardless of how that ranking is performed, captures the essence of VS ([1], quoting [2]). The choice of ranking algorithm generally depends on the information known on the target, knowledge of compounds active in the relevant biological assay, how dissimilar the desired ligands need to be from known bioactive molecules, and what percentage of the ranked database would be selected for experimental testing. The smaller the percentage of compounds to be tested, the more efficient the ranking algorithm needs to be to result in successful hits from VS.

The most common VS method is a similarity-based (almost always executed through the use of a fingerprint) or substructure-based search. These are so integrated into medicinal chemistry practice that they are often overlooked as being amongst the most common and effective VS methods. However, given one or more active compounds, chemists invariably attempt to identify similar molecules using substructure and similarity queries. Substructure and similarity searching is often the unexciting but highly effective follow-up of a more complex virtual screen that attempts to find new lead matter. Even when only a few analogs turn up as initial hits, substantial structure–activity relationships of an entire series can sometimes be gathered without requiring new synthesis. The continuing interest in fingerprint-based methods is covered in more depth in the LBVS section of this review and in several reviews in the literature [3–7].

The more challenging scenario arises when the need to identify new scaffolds or series becomes the driver of the VS experiment. Often, the ratio of the number of compounds selected for testing to the size of the database of compounds screened, SSR (selection to superset ratio), is in the range of a thousandth or less. Success (which could be 1% or more of selected compounds having relevant biological activity) while selecting in such low SSR situations (small number of molecules selected from a very large collection) has won VS the recognition as a distinct function of computational chemistry that can deliver new leads to a drug discovery effort complementing experimental methods like high-throughput screening (HTS). Mostly such VS is done to select compounds from databases typically present in medium to large pharmaceutical companies or compendia of commercially available compounds or combinatorially synthesized collections provided by vendors or combinations thereof. A characteristic of such databases is the variable extent to which different segments of chemical space are over or under represented.

VS methods to identify new chemical series can be broadly classified into three classes:

1. Methods that rank compounds based on some measure of similarity to known actives, based on 2D or 3D structure of the molecule (LBVS).
2. Methods that deduce a pharmacophore, an arrangement in 3D space of features that contribute or detract from binding and look for its presence in the database that is searched. This method places emphasis on features like hydrogen bond donors, hydrogen bond acceptors, acidic or basic units and hydrophobic fragments and opens the possibility of identifying unexpected scaffolds with required features (pharmacophore-based VS or PHBVS).
3. Methods that utilize structural data of the target, generally identified by protein crystallography, to look for molecules that complement the "binding site" through favorable protein–ligand interactions (protein structure-based VS or SBVS).

The choice of method used is often facilitated or constrained by the information available. In the absence of structural information on target, if one or more active small molecules are known, LBVS or PHBVS are feasible. If no active compounds are known, but an experimental or computational model of the protein structure is available, SBVS can be considered. If both active compounds and target structure are available, one or more appropriate methods can be applied, or multiple methods combined.

There have been a number of very helpful reviews of aspects of VS in the past few years. These have focused on either specific methods, or on the field as a whole. Cramer has provided an interesting review of methods of lead-hopping, concentrating on technologies applicable to find scaffolds very different from the initial scaffold known to be active [8]. This is especially useful if there is some doubt as to whether a molecule from the current series can be developed with sufficient chemical novelty to allow it to be patented. Jalaie and Shanmugasundaram have reviewed the state of the art prior to 2005 [9] as have Reddy et al. [10]. A review focused on LBVS has been published by Hert et al. [11]. Finally, a broad and characteristically trenchant review has been provided by Klebe [12]. In this review we will focus on advances and successes reported in the past 2 years, with the perspective of practitioners of the art in two large pharmaceutical companies.

1.1 Benchmarking Virtual Screening Methods

Numerous researchers in academia and industry have worked to advance the performance of VS methods. Many sets containing molecules active at a given target mixed in with known or presumed inactives (better referenced as decoys) have been created and have been used to demonstrate the performance of individual methods, or compare the performance of multiple methods. Table 1 provides a summary of many of these data sets, most of which are publicly available. A key consideration is the choice of inactives/decoys present in these datasets. Ideally, the physicochemical profile of the inactives/decoys should be matched to those of the actives, thereby preventing the observed enrichment from being a surrogate for property differences between active and inactive members. This is a consideration because many scoring functions are somewhat correlated with the molecular weight and lipophilicity of the ligands docked and scored.

Generally, performance of a method is often judged in one of two ways. The first is the *enrichment factor*, *enrichment* for short, which is the ratio of the cumulative number of actives in the top N% of the total number in the dataset to random retrieval rate. Many early studies focused on the enrichment obtained when the top 10% of the dataset was screened. However, this is operationally unrealistic if compound collections exceed 100,000 compounds, which is common in mid- to large-sized companies. A more realistic test is the enrichment obtained in the top

Table 1 Reference data sets, and location as of 2008

Data set(s)	Actives	Link
Cox2	128	http://www.ncbi.nlm.nih.gov
Estrogen receptor	55	
Gyrase B	55	
Neuramidase	83	
P38 kinase	55	
Thrombin	67	
DHFR	100	
Factor Xa	100	
ZINC	Variable	http://zinc.docking.org/
DUD	2,539 actives against 40 different targets	http://dud.docking.org/

0.1–1% of the compounds ranked. Many papers now include 10%, 5%, and 1%. This method unfortunately depends on the ratio of actives to decoys present in the dataset and makes comparisons across datasets difficult.

Another measure, that is independent of the ratio of actives to decoys, is the more comprehensive receiver operating curve and an enhanced version [13, 14] and reduces the dependence of the success measure on the number of decoys in the set. This also graphically demonstrates the enrichment as a continuum, plotting the fraction of the actives retrieved (true positive retrieval) against the fraction of the inactives retrieved (false positive retrieval) [15]. The first ratio is the sensitivity of the method (fraction of compounds that are predicted to be active out of the total true actives present in the sample) and the second is the specificity of the method (the fraction obtained by subtracting from 1 the ratio of the true negatives to total negatives which would be the ratio of compounds falsely predicted to be positive out of all the inactives). The area under the curve, AUROC, is a measure of the efficiency of the method. As the VS method gets better the area under the ROC curve will approach 1. The method is better than random if the area is >0.5 and this method allows comparison across datasets since the curve shape and area are independent of the size of the dataset. Figure 2 shows a ROC curve reproduced from a PubMedCentral Article, which also has a lucid account of this widely used statistical method [16, 17].

The variety of test data sets has helped to broaden our understanding of how different methods perform under a variety of circumstances. Notably, the enrichment

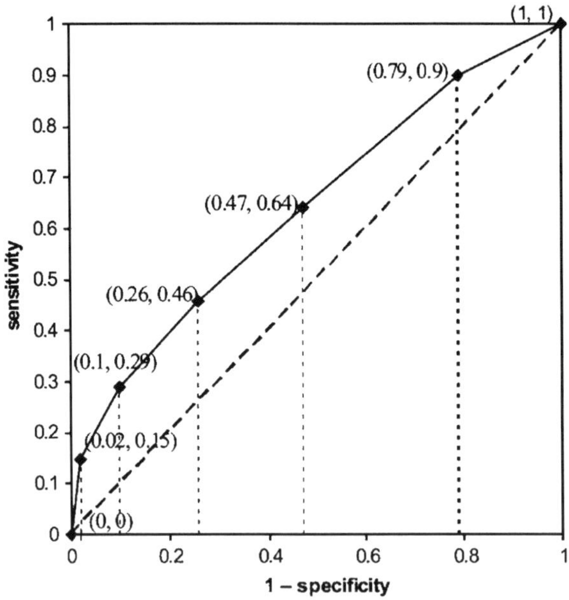

Fig. 2 A receiver operator characteristic curve reproduced from PubMedCentral http://www.pubmedcentral.nih.gov/articlerender.fcgi?artid=1065080 and [16]

in structure-based methods that utilize docking and scoring can be highly dependent on the quality of the scoring function employed and how sensitive it is to small errors which creep in at the docking step [18, 19]. Therefore, by testing performance against a variety of targets, a more realistic assessment of a given technique can be produced. This information allows the experienced drug hunter to better tailor their VS experiment to the protein class. In recent years, there has been a significant advance in the ease of availability of curated datasets publicly available ZINC and DUD being notable examples [19, 20] that allow comparison of the performance of different methods using active/decoy combinations available to all practitioners of the art.

A key issue that arises in practice, is whether the experiment is biased toward enrichment (most actives for the number of compounds screened) or novelty (compounds identified are drawn from different chemical classes). Generally, any protocol leading to a small number of compounds tested is biased toward enrichment. In this case, the strategy can often involve multiple serial VS methods with aggressive property filtering. With fewer experimental constraints, the computational chemist is afforded the luxury of fewer assumptions. For example, by performing a pharmacophore-based VS, but not refining the data set by docking into a known protein complex, compounds that might otherwise be excluded due to a change in the active site might be found. Furthermore, the practice of inspection generally aids in eliminating unrealistic binding modes or undesirable chemical functionality [21]. This is done with the risk that the unexpected turns out to be true.

1.2 Database Creation

One big advantage of the VS experiment is that the compound screened need not be available physically and possibly not exist at all. Generally, the enrichment available through VS is thought to be insufficient alone to justify de novo synthesis. An exception to this rule is the practice of computationally screening a large combinatorial virtual library, identifying potential actives and following up with combinatorial synthesis of a smaller subset of the library [22–24]. As the enrichment offered by VS methods improves and with sufficient synthetic capacity, this is an assumption that could be challenged.

There are a number of commercial vendors who have made their compound collections available in formats amenable to translation into databases that can be screened with relevant software. Generally, starting from any standard format (MDL Mol or SDF, SMILES), compounds can be converted into a format required for database searching. For docking or 3D database searching, this also requires the creation of one or more 3D conformations of the molecule, which are stored for using by the screening software. Software such as CORINA [25], CONCORD [26], OMEGA [27] and Catalyst [28] have procedures to convert 2D to 3D coordinates and to generate a family of minimized conformers.

One element of database generation that is a key consideration is whether to expand the representative compounds to include alternative tautomers, protonated and deprotonated forms of the molecule, and also to enumerate stereochemistry fully if not specified in the input. Depending on the molecules in question and the options considered, these can lead to a 10-fold increase in the size of the database to be explored. However, such an expansion is necessary if methods are used that are sensitive to such chemical precision (e.g., docking). For 3D similarity searching, it is sometimes more efficient to consider various modifications to the query, leading to multiple searches against a smaller database.

A further consideration when combining databases from multiple suppliers is how to identify and deal with redundant compounds. Here, some method of mapping multiple supplier information onto a single compound is needed for efficiency. Generally, all information can be mapped, although some consideration of cost and supplier reliability may allow a hierarchy of supplier information to be applied.

1.3 Database Filtering

For many practical reasons, some element of filtering is often applied either at the point of creating a subset of "chosen" compounds for VS, or to a VS hit list before ordering compounds for testing. A number of property-based filtering criteria are available. By far the most famous are the Lipinski rule of five criteria [29–31]. By reviewing the computed properties of known oral drugs, a pattern emerged that suggested that an orally absorbed drug had a molecular weight less than 500, fewer than five hydrogen bond donating groups, fewer than 10 hydrogen bond accepting groups, and a clogP less than five. Veber et al. [32] further proposed that intestinal permeability decreased as the polar surface area (PSA) exceeded 140 Å^2. Inspired by these studies, a number of researchers demonstrated that the impetus of most lead optimization efforts tended to add size and lipophilicity to the molecule, and that the desired lead should be smaller and less hydrophobic than the eventual drug [33, 34]. This work has caused a number of researchers to limit the size of the ligands that are introduced into the "collections" of potential leads. From a practical perspective, these rules have considerable operational benefit, because they limit the size and conformational complexity of the molecules assessed. This leads to smaller databases, and less search time per molecule. Another filter which addresses size complexity from an operational perspective is the number of rotational bonds present in the molecule [35]. Prefiltering the collection to remove molecules with more than 10 rotatable bonds is a common practice except in specialized situations where there is prior knowledge that a long linear chain is a necessity for biological activity.

Another assessment of drug-likeness is afforded by Jorgensen's Qikprop family of ADME models [36]. In addition to the individual predictions, Jorgensen has proposed a rule-of-three. This proposes that a successful drug will have predicted solubility (log *S*) >−5.7, predicted *p*CaCO >22 nm s^{-1}, and less than seven predicted primary metabolites. Qikprop will also provide an assessment of whether the compound is considered similar to molecules in the training set for these models. Given the uncertainties about applying any general model to a diverse set of molecules, this might be a consideration later in the assessment phase of the VS. It might also be used to prioritize the eventual experimental hits for experimental ADME assessment.

Additional filtering options arise from considerations of potential toxicity. Davis et al. [37] published an extensive list of chemical fragments that were proposed by medicinal chemists to be either reactive or that might be linked to toxicity. Such filters can remove unwanted or suspect functionality prior to testing to increase the likelihood of a hit being attractive as a chemical lead. Hit lists can also be filtered by any number of general QSAR models of ADME properties. While effective at further reducing the numbers of compounds in a database or list of virtual hits, the applicability of a general model on a compound from a series on which the model was not trained is suspect, at best. Such models are best applied late in the process, when some critical assessment of their validity might be attempted based on known data or by comparison of the compounds in the hitlist to the training set of the model.

2 Ligand-Based Methods

2.1 Introduction

Probably the most efficient ligand-based search method devised to date is the similarity search based on chemical fingerprints. There is a wide range of ways of defining "features" that can be mapped as part of a fingerprint: atoms, atom pairs, chemical functional group fragments and connected bond fragments [5, 38, 39]. These can then be further generalized, either by atomic properties, atom type, interactions afforded by the chemical features (e.g., hydrogen bond donor/acceptor/both), or various topological and graph theory indices [40, 41]. The choice of information encoded and the degree of generalization or abstraction can be tuned in an attempt to bias the "similarity" to match molecules with desired common attributes.

Clearly, within the conceptual framework described above, there is extensive room for exploration in creating fingerprints and similarity measures to retrieve molecules based on varying conceptions of "similarity" [42–44]. The simplest types of fingerprint consist simply of features indices that map the presence or absence of a small library of functional groups. The most well known and effective are the MACCS keys. These were initially chemical feature indices, that we later used successfully as a similarity metric.

A richer fingerprint description is provided by the Daylight [45, 46] or UNITY (Tripos Inc., St. Louis) fingerprints. These incorporate a much broader range of features, notably including connected bond path fragments up to seven bonds long.

Additional commonly-used fingerprints offer alternative ways to encode path lengths. The ECFP [47, 48] series of fingerprints used in Pipeline Pilot use a different algorithm to code path lengths of four bonds (ECFP4) or six bonds (ECFP6) or higher in length. If the atoms are genericized to a small number of roles (e.g., hydrogen bond donors and acceptors), the topologically related family of FCFP fingerprints [49] can be generated. These fingerprints have proven useful in multiple roles including similarity searching, complexity analysis, and QSAR model generation using Bayesian learning machines [50].

Another family of fingerprints available are the MOE pharmacophore fingerprints accessible through software from the Chemical Computing group [51]. In this system, the atoms are generalized into a smaller vocabulary of pharmacophore features, after which the fingerprint is constructed based on connected paths.

Feature-based fingerprints should be noted for their inclusion of pharmacophore feature types, and counts along with structural and property data into a single fingerprint for VS [52–56]. One of these arises from the Leadscope hierarchical classification of 64,000 scaffolds which has been converted into a fingerprint and used in similarity analysis [57]. Another more customizable set is one put together by Digital Chemistry software [58]. Here, a wide number of feature, path, and generalized features can be created as a huge dictionary, and then a subset of bits with the best characteristics for a given task can be chosen. Unlike many folded fingerprints, this approach has the dual advantages of being able to tailor the fingerprint to the task, and to map back the features set to the molecule.

The pragmatic beauty of the chemical fingerprint is that the more common features of two molecules that there are, the more common bits are set. The mathematic approach used to translate the fingerprint comparison data into a measure of similarity tunes the molecular comparison [5]. The Tanimoto similarity index works well when a relatively sparse fingerprint is used and when the molecules to be compared are broadly comparable in size and complexity [5]. If the nature of the molecules or the comparison desired is not adequately met by the Tanimoto index, multiple other indices are available to the researcher. For example, the Daylight software offers the user over ten similarity metrics, and the Pipeline Pilot as distributed offers at least three. Some of these metrics (e.g., Tversky, Cosine) offer better behavior if the query molecule is significantly smaller than the molecule compared to it.

When used in the VS context, the fingerprints of both query molecules and the database of molecules probed must all be computed. Generally, the fingerprints of the database compounds are often precomputed and held as additional attributes for each molecule. For each type of fingerprint and similarity metric, some similarity threshold is often applied to limit the number of hits achieved. Because of a fair amount of work early in the 1990s, the 85% similarity threshold is often applied ($T_c = 0.85$). However, this was first done in the context of Daylight fingerprints and Tanimoto indices, and should not be extended to other systems without further validation. For example, our own work suggests that similar molecules will still be retrieved using a 70% similarity threshold with UNITY fingerprints. Researchers at Leadscope [59] applied a 45% similarity threshold to comparisons using their proprietary fingerprints. Some validation is generally needed when considering a new fingerprint and similarity metric combination.

A different approach to molecular similarity is offered by various descriptor sets generated either from calculated physical properties (e.g., molecular weight, cLogP) or more complex metrics derived from graph theory. An example of the latter are BCUT descriptors developed by Dr. Robert Pearlman [60, 61]. This is currently available as part of DiverseSolutions (Tripos Inc., St. Louis). These descriptors are generally understood to encode the molecular hydrogen bond donating or accepting nature, charge, or polarizability. Operationally, this metric has the advantage of scaffold hopping in practice [62–64]. A variant of this approach is available from the CCG MOE software as QSAR descriptors [51].

2.2 Case Studies

Given the relative simplicity of ligand-based methods, it is interesting to note that in only comparatively few published reports of VS successes do the authors rely primarily on ligand-based methods. Of these studies, most appear to combine an interest in a given target with an interest in providing proof-of-concept for some extension of chemoinformatic theory.

In Table 2 we highlight pertinent information from a number of studies. We did not aim to be exhaustive, but rather to provide enough examples to provide a flavor for the type of studies performed. Of the studies in Table 2, one element to note is the small number of compounds tested in five cases. Despite starting with databases that range from 37 K to 2.5 million compounds, most researchers end up actually testing less than 100 in most cases and several hundreds at most.

An example of the value of VS based on descriptors alone is that of the identification of inhibitors of 5-lipoxygenase by Franke et al. [73]. 5-Lipoxygenase catalyzes the first transformation of arachidonic acid to leukotrienes that mediates many inflammatory responses. It has also been proposed as a contributor to atherosclerosis, cancer and osteoporosis. To seed their study, 43 known 5-lipoxygenase pathway inhibitors were used. The investigators chose the AnalytiCon Megx library of purified natural products as their database, which contained 1,298 compounds at the time of testing and the Nat-X library containing 7,839 compounds [74]. The CATS-2D topological pharmacophore-pair descriptors [75, 76] were used, and 430 hits (10/query) were assessed visually for the novelty of their scaffolds. Just 18 were tested, of which two showed activity in a cell-based assay. Both hits were from a library of natural products derived from α-santonin. For each hit, several close neighbors were selected for screening from the NAT-5 library. Additional hits were obtained for both series. Since cell-based screening was performed first, followed by receptor-based screens, some ambiguity remains as to whether the hits – especially related to series 2, are genuine 5-lipoxygenase inhibitors or act elsewhere in the pathway probed by the cellular assay.

Table 2 Examples of ligand-based VS workflows

Target	Notes	Outcome	Reference
Dopamine D2, D3	SPECS db (230 K), NN, clustering, SOM	9 D3 antagonists, 6 D2 antagonists of 190 tested	[65]
Kir6.2/SUR1 K ATP channels	ZINC db (65 K), FLAP screening to 1,913 compounds	3 hits of 32 tested	[66, 67]
L-type calcium channels (voltage-gated calcium channels L-subtype)	Similarity to Diltiazem and a second ligand. ZINC db (~50 K commercially available subset screened but most filtered to achieve desired PK profile using VolSurf). SHOP similarity, and feature-presence filtering down to 36 compounds	7 hits 18 tested. active in a vasorelaxant assay and some had novel structures.	[67]
5-Lipoxygenase	AnalytiCon Discovery db, Similarity based on 2D CATS descriptors	18 hits/430 tested	[68]
ICAM-1/LFA-1	Database of 2,500 K, custom minifingerprints based on pharmacophore pairs	1 hit/25 tested	[69]
Na/K ATPase	ICB natural product database (37 K), QSAR, Chemfinder similarity search based on ouabain	4 hits/10 tested	[70]
PDE1, PDE5	SPECS (88 K), CART regression trees based on 2-point pharmacophores	7 hits/19 tested	[71]
Mycobacterium tuberculosis	Recursive partitioning, similarity to conceptual virtual libraries	1 hit/4 tested	[72]

A second example of a VS exercise that was largely fingerprint-based was that of Boecker et al., in search of novel series for dopamine D2 and dopamine D3 blockers [65]. A set of known actives consisting of 472 dopamine D2 and D3 ligands was assembled from the literature. The SPECS database of 230,000 compounds was chosen from which to identify compounds. Two descriptor sets were calculated: MOE2D [51] and CATS3D [77] for both query and database molecules. Neighbors of the known actives were then identified using NIPALSTREE hierarchical clustering, hierarchical k-Nearest Neighbor analysis, and a self-organizing map analysis. These analyses yielded 37, 144, and 52 neighbors respectively. These hitlists were culled by considering druglike properties, the presence of an ionizable nitrogen (a key pharmacophoric element) and novelty. Of 17 compounds eventually purchased and tested, nine had potent (K_i <1 μM) D2 binding and six had potent D2 binding. The most interesting had dopamine D3 binding of 65 nM and was 13-fold selective over D2. As a follow-on study, a pharmacophore model was built using the MOE [51] software and the dataset of literature and recently identified molecules. This was applied to the SPECS database, and four additional compounds were ordered. The best had dopamine D3 binding of 65 nM and was mildly selective over D2. All four had binding of <10 μM at either the D2 or D3 receptors.

An interesting example of the use of novel fingerprints developed by the cheminformatics group at the University of Perugia and marketed by Molecular Discovery Inc. is afforded by a paper describing the search for novel potassium channel openers reported by Carosati et al. [66]. Compounds that open pancreatic ATP-dependent potassium channels may help regulate insulin secretion in diabetes. The ZINC database [16, 19] of 65,208 compounds (in 2005) was reduced to 1,913 compounds by applying pharmacokinetic filters. Molecular weight was restricted to between 200 and 600 amu, and clogP to between 1 and 5. In addition, three Volsurf [78] ligand-based models were applied to select compounds predicted to have good absorption, limited blood-brain barrier penetration, and adequate cell permeation. From this smaller pool of compounds, molecules were chosen that were similar to six known potassium channel openers. This was accomplished by principal components analysis of the GRIND [79] (grid-independent pharmacophore descriptors), multivariate similarity of TOPP [80] (three-point pharmacophore-based fingerprints) and pairwise superposition and scoring of FLAP [80] (four point pharmacophore-based descriptors) were calculated for query and target molecules and similar compounds identified. After inspection and selection, 3 compounds of 32 eventually purchased demonstrated E_{max} >100% when tested in channel preparations. The paper highlights that each different type of descriptor used identified different compounds, which were combined into the final set that was ordered.

A fourth example highlights the value of generating a predictive model of activity from known SAR and then applying this model to a database of compounds. Yamazaki and coworkers undertook this analysis to identify new classes of PDE1 and PDE5 inhibitors for development as potential cardiovascular therapeutics. Existing SAR for 130 compounds was initially used to train a CART recursive partitioning model, with 10,000 diverse compounds selected from 88,000 SPECS compounds used as an inactive background. One hundred and sixty eight descriptors were calculated based on binned distances between pharmacophore pairs, and an additional 12 physical property descriptors were added. The SPECS database of compounds was searched using the derived model, although filtered to ca. 50,000 compounds by comparing the latest version of the catalogue with a 1998 version and removing common (older) compounds. This was done to bias the compounds to those likely to be available. One thousand eight hundred and twenty one putative inhibitors were identified using the CART model, of which 100 were selected by diversity analysis. From these, 19 compounds were tested, of which 11 showed >50% at 10 μM and 7 were of interest as dual PDE1 and PDE5 inhibitors.

3 Pharmacophore-Based Methods

3.1 Introduction to Methods

The notion of the "pharmacophore" has a long and successful tradition within medicinal chemistry. Before the visualization of protein–ligand interaction brought on by crystal structures, chemists working within a given series would – by trial and error – identify those parts of the molecule most associated with a desired biological activity [81-83]. Provided the pharmacophore remained constant, changes elsewhere in the molecule might modulate activity but often ensured that potency was retained with exceptions arising only when additional molecular fragments caused serious disruption. This idea can be further generalized; if a pharmacophore is satisfied by other functional groups, or by comparable groups or atoms arranged in a spatially comparable way on another scaffold, then the two classes of molecules might share similar biological activity. This precept – that even when 2D topology might not suggest a common pattern of features, the presence of required pharmacophoric elements in desired spatial geometry is sufficient to provide relevant biological activity – has powered and continues to power the contributions of computational modeling and VS to drug discovery and design, and is well reviewed in the literature and a few are included [84–89]. These ideas were then extended to searching a database of 3D structures for ligands that matched a 3D pharmacophore [90, 91]. These are the methods that are generally referred to by the "pharmacophore-based VS" shorthand. Implicit in some of the discussion about pharmacophore-based fingerprints above is that another use of the term "pharmacophore" is for any scheme that refers to a collection several atoms or functional groups to pharmacophore features without the 3D geometry being included. However, in this section, we will tend to focus on methods and case studies in which a 3D pharmacophore method was applied.

The 3D pharmacophore, in its simplest form, is the presence and geometric arrangement of a combination key elements, usually selected from hydrogen bond donor/s, hydrogen bond acceptor/s, aromatic ring/s, and hydrophobic group/s. In the absence of 3D structures of receptors complexed to ligands, the pharmacophore was considered the major biologically relevant metric [88] that related molecular structure to biological activity. However, as one could easily perceive, a collection of descriptors, which capture the characteristic elements, the charges, hydrophobic character and shape, can readily describe a 3D pharmacophore in finer detail. Such descriptors were deployed in modeling and design under the general umbrella of 3D QSAR and VS experiments were accomplished with a spectrum of variations that ranged from a simple collection of pharmacophoric binding elements to multidimensional QSAR. These have been covered in many recent and almost recent reviews and we include a large selection of them for the benefit of the reader who wishes to explore applying those methods [7, 8, 92–103].

A number of very useful tools have emerged using methods that rely on shape matching or surface similarity matching. These include the ROCS method from Open Eye (www.eyesopen.com, [1]) and the Surflex-Sim [104] surface-matching method developed by Jain and currently marketed through Tripos. The shape-based method from Open Eye, called ROCS [105] has emerged as a frequently used tool in the hands of industrial chemists [1]. ROCS relies on the conversion of a single molecule in a putative bioactive conformation into a series of Gaussian grid functions representing shape or atomic character. This probe is compared to similar information coming from a precomputed database of stored conformations, and a scaled similarity function is generated from either shape overlap or similarity of atomic character. Recent publications highlight the need to employ both types of information to ensure enriched screening lists [106]. This method is distinguished by its speed, reasonably simple command-line interface, parallelization, and robust behavior across multiple ligand classes.

The Surflex-Sim method operates significantly differently [104]. Each of the molecules is surrounded by a set of "observer" points that characterizes the local character of the surface and potential interactions. Two similar molecules will have a common subset of comparable observer points. A optimal alignment occurs when the differences in pharmacophore character and molecular surface inferred from the observer points are minimized between two molecules. To speed up the algorithm, large molecules can be fragmented into parts which are then compared, and then tested for consistency. This feature also makes the method capable of identifying alignments when one molecule is much smaller than the other.

An older but effective and widely used method is the Catalyst program from Accelrys. Like ROCS, it operates as a VS tool against a database that contains a precomputed conformational expansion for all ligands. Multiple conformations of every compound are stored. It is distinguished by the ability to generate a 3D pharmacophore based on hydrogen bond donating and accepting elements, hydrophobes, and optionally positively and negatively ionizable functional groups. If trained on known ligands with three or more orders of magnitude of biological data, a robust activity prediction equation can often be generated. This function can be used as a scoring function in the subsequent VS experiment. Unlike a similarity function, this type of function can penalize for features that are already known to detract from biological activity. However, in the absence of such a scoring function, Catalyst can operate in a similarity scoring mode. Effective variations of Catalyst like functionality are also available from Computational Chemistry software from Chemical Computing Group and Schrodinger.

A third and slightly older method available is the UNITY package from Tripos Inc. This also relies on the user to identify pharmacophore features and spatial arrangement. When multiple compounds and biological activity is known, this can be used to focus on a limited number of features or to exclude specific volume from the molecule. The compounds in the database are then compared to the query pharmacophore using a flexible directed tweak algorithm. In practice, some tuning of tolerances and features are often necessary to achieve reasonable recall of actives. Validation with known actives against a small, diverse background of inactives is often recommended prior to a large-scale database search.

A complementary method that derives pharmacophores from a protein crystallographic complex is Ligand Scout from Inte:Ligand, [107] (www.inteligand.com). This method has a limited vocabulary of pharmacophore features that includes hydrogen bond donors and acceptors (and extension points), normals to aromatic rings, and hydrophobes. In practice, it has been used to convert the putative or known binding sites into pharmacophore search queries, after which the pharmacophore information is transferred to software such as Catalyst or MOE. In validation studies, it is effective at reproducing relevant binding modes.

Most of the pharmacophore methods employ a set of features that include hydrogen bond donors and acceptors, hydrophobic volume, sometimes excluded volume, and also positive and negative ionizable groups. An alternative pharmacophore description is that of the Cresset software [108, 109] [www.cressetbmd.com]. This software relies on using the extrema of the electrostatic potential, as well as a description of hydrophobic regions, to create a database query. To improve the quality of the electrostatic potential around the molecule, additional charge-bearing features are included in the force field representation to reproduce delocalized pi electrons better. The field pattern is then compared to a database of precomputed field representations based on multiple conformations for each molecule. The software offers options to generate a consensus pharmacophore from multiple ligands and to align the molecules retrieve for visual inspection.

3.2 Case Studies

A number of recent examples of the use of pharmacophores as a primary VS method have appeared in the literature. Table 3 provides a selection of these studies, with outcomes listed. The databases searched range in size from 630 molecules to 1.7 million molecules. Of the studies shown in Table 3, the Catalyst software is the method most often used, followed by UNITY and the FlexS superposition tool.

An excellent example of the ability of pharmacophore methods to search a large database rapidly is afforded by the VS done by Schuster [110] and coworkers to find antagonists of 11-β HSD. This enzyme catalyzes the conversion of 11 ketosteroids to 11-β hydroxysteroids. Inhibition of glucocorticoid overexpression may be effective in treating metabolic syndrome, and inhibition may also have a role in treating diabetes and muscle atrophy. Known selective 11-β HSD1 inhibitors were used to train a model in Catalyst that contained a donor location, an acceptor location, and four hydrophobes. The ability to retrieve inhibitors was tested both using the known inhibitors against a random set of molecules (presumed inactive) and from the WDI database of approximately 63,000 compounds. A second model was generated from inhibitors that bound to both 11-β HSD1 and 11-β HSD2 and was tested in a similar way.

Table 3 Examples of VS using primarily pharmacophore methods

Target	Notes	Outcomes	Reference
11-β HSD	Database of 1,700 K, Catalyst –>31 hits	7 hits/30 tested	[110]
AR downregulating agents (ARDA)	Maybridge (60 K) and NCI (239 K) Catalyst –>41 hits	6 hits of 17 tested	[111]
Alzheimer's tau protein	Maybridge database, 136 identified	2 hits of 19 tested	[112]
CoX-2	Maybridge database (12.5 K) Catalyst search followed by GOLD docking	5 hits of 8 tested	[113]
Chloroquine-resistance reversal agents	3D QSAR		[114]
Fetal Hb transcription inducers	TFIT pseudoreceptor, Similarity search of 630 candidate molecules	2 of 26 active	[115]
Ginkgolides as GABA modulators	Pharmacophore search of 300 K structures	No hits of 31 tested	[116]
GR-Glucocorticoid receptor	Commercial db (718 K) filtered to 862, searched by FlexS	One series	[117]
Chalcones	Chemical library probed with pharmacophore	Ligands active in vitro and in vivo	[118]
Mycobacterium tuberculosis H37Rv	Pharmacophore selection of 95 compounds from a database of 15 K, docking to further reduce candidates	4 potent hits	[119]
PPARγ	Maybridge db (62 K), Catalyst	Novel series	[120]
Pfmrk: plasmodium falciparum	3D QSAR		[121]
SIRT-2	Maybridge, Leadquest dbs, UNITY search	4 of 11 tested	[122]
T-type calcium channel	Maybridge (55 K) and ion channel inhibitor db (8 K), Catalyst search	3 hits of 25 tested	[123]

The 2 pharmacophore models were used to search a database of about 1.8 million compounds assembled from 12 commercial databases. Hypothesis 1 returned approximately 20,000 hits, which were aggressively filtered using the Catalyst scoring function, lack of hit to a hERG pharmacophore, clogP <5, fewer than five donors and ten acceptors. Fifteen compounds remained and were available for purchase after filtering. The second hypothesis returned 107 hits, of which 15 were chosen for testing. Seven of the 30 compounds eventually purchased inhibited the activity of cell lysates by at least 70% at 10 μM.

A second study points out the need to develop a strategy consistent with the computational tools being used. Ray et al. [117] performed VS based on 3D similarity to three glucocorticoid receptor blockers. High glucocorticoid levels may be linked to the psychotic symptoms of psychotic major depression. A commercial database of 718,000 compounds was aggressively clustered and filtered to 862 compounds. FlexS [124], a 3D similarity program, was then used to assess the similarity of these molecules to three known glucocorticoid receptor blockers. The filtering was needed as FlexS performs a flexible superposition and is comparatively slow. Because one of the query compounds was racemic, both enantiomers were built and used as query molecules. Conformational searches of the query molecules identified low energy conformations for each. From these searches, 123 compounds were identified, which were further narrowed to 18 by inspection and supplier considerations. Of these, one compound was reported to block the glucocorticoid receptor with a K_i of 4.5 μM. Two rounds of similarity searching identified more potent analogues, the best of which had a K_i of 16 nM in in vitro screening. This demonstrates the need to match the database to the computational capacity availability, the implicit value of inspection, and the value of follow-up similarity searching to rapidly fill out SAR.

A third study demonstrates the value of using a pharmacophore obtained from the binding site of a protein complex. Tervo and coworkers [122] used the UNITY software to create two pharmacophore hypotheses based on the docking of three known sirtuin-2 histone deacetylase inhibitors. Sirtuin-2 is believed to be essential to the mitosis of some cells and may play a role in fat storage, some cancers, and possibly Alzheimer's disease. Based on the docked poses, a pharmacophore containing two hydrophobic locations, a donor atom, and one of two possible acceptor atom sites was defined, as well as regions of excluded volume. Lipinski filtering was applied, with the donor atom limit reduced to three and the acceptor atom limit reduced to seven. Flexible searches of the Maybridge and LeadQuest libraries were performed, which resulted in 34 compounds. These were reduced to 32 compounds by applying the Volsurf [125] permeability model. Further inspection led to the purchase of 11 molecules. Of these, four showed IC50 inhibition of <200 μM in in vitro testing.

4 Receptor Structure-Based Methods (SBVS)

4.1 Introduction to Methods

The genomic era continues to transform itself into the proteomic era [126]. A number of entities ranging from pharmaceutical companies to publicly funded academic research groups have been solving the crystal structures of many genes, and tackling ever more complex crystallographic challenges [127, 128]. For many families of drug targets there is now one or more crystal structures available of the target itself or a close homolog or ortholog.

The elegance and promise of the availability of structure for ligand discovery – that once we have an apo site in atomic resolution, we can find molecules that bind tightly to it by generating a very large number of virtual complexes, followed by scoring, ranking and selecting the very best – has been a holy grail of structure-based discovery ever since Irwin (Tack) Kuntz and colleagues came up with a program called Dock roughly two decades ago [129] that could identify molecules from the Cambridge Crystallographic Database that could fill a given protein site. Much has happened in the last two decades and a recent review, interestingly with the same researcher being the first author, gives a picture of the state of the art [130]. In the intervening 20 or so years, at least 50 docking programs and their variants have been developed. Docking has come of age and docking software available can in most cases reliably and quickly reproduce observed crystallographic binding modes of protein–ligand complexes with RMS variations approaching the experimental error in the crystallographic experiment that characterized them. With robust docking tools and fast, cheap and plentiful computing power, it is a surprise that SBVS has not replaced experimental screening. In practice, however, several published and unpublished success stories notwithstanding, this still stays a challenge, to the point that successful SBVS is not as routine as one could have expected it to be per our outlook from a decade ago [131]. This is despite vigorous development of docking methods and scoring functions by the computational chemistry community for well over a decade chronicled in the representative set of citations here [21, 36, 132–142].

Part of this disconnect between expectations and performance in SBVS originates from the way protein–ligand interactions are quantitated to arrive at selecting the best pose of the small molecule in the receptor site rapidly, or the way the "docking problem is solved." These make approximations in correctly describing the entropy change upon binding, and free energy components such as free energy of solvation, in order to sample and evaluate rapidly a substantial number of conformations including multiple poses for each conformation of the small molecule in the receptor, assuming the receptor is held rigid, which is common in SBVS applications. When applied to the problem of choosing the correct docking pose for the same molecule, the changes in solvation and entropy tend to become negligible from pose to pose, leading to substantial success in selecting the best pose from amongst a set likely of the docked poses. This is only a generalization and can be influenced by the nature of the binding site, in terms of whether it is a site that deviates to the extremes of hydrophobic or hydrophilic character, whether there is potential for less or more protein movement, and whether the ligand in question has more or less rotatable bonds [135] resulting in one or more docking programs being better than another for a particular combination of receptor and ligand. In addition, most docking programs generate a series of poses that are relatively closely spaced in their "docking score", the pseudoenergy function used by the docking software

Table 4 Virtual Screening examples using primarily SBVS

Target	Database	Protocol	Outcome	Reference
E. coli primase	Commercial (500 K)	Glide, filtering	4 hits of 68 tested	[143]
RNA methyltransferase	SPECS, Maybridge (300 K)	FlexX, filtering and inspection	2 series, 8 hits of 33 tested	[23]
LXR	Proprietary 135 K	Glide	1,295 tested, one hit series highlighted	[144]
S-Adenosylmethionine carboxylase	NCI diversity (2 K)	Glide	18/133 tested	[145]
Sex-hormone binding globulin	90 K, 52 K remaining after filters applied	Glide, top 16 tested	4 hits of 16 tested	[146]
CYP 2D6	6 K	GOLD,	11 hits of 16 tested	[147]
Falcipain 2 and 3	355 K	GOLD	22 of 100 tested	[148]
SARS CoV	361 K	EuDOC	1 of 12 tested	[149]
Chk1 kinase	700 K	RDOCK	9 hits	[150]
Heme oxygenase	Commercial databases (800 K)	DOCK	8 hits of 27 tested	[151]
Ubiquitin C-terminal hydrolase	Chembridge (33 K)	DOCK/GOLD	3 hits	[152]
12-Lipoxygenase and 15-Lipoxygenase	Chembridge diversity (50 K)	GLIDE SP, XP	3 hits of 20 tested	[153]
ER alpha	SPECS 202 K	LIGIN, Chemgauss	3 hits of 7 tested	[154]
NkkB	Proprietary 5,000 K	4SCAN/ProPose	1 hit shown of 236 tested	[155]
Cyclophilin A	SPECS 280 K, 85 K after filtering	DOCK, CSCORE, FlexX	15 hits of 82 tested	[156]
Potassium K+ channel	DOCK 4.0/homology model	ACD (200 K)	6 hits of 20 tested	[157]
HCV 3D polymerase NS5B	GLIDE	2,000 K commercial	1 series from 50 tested	[158]
SARS Coronavirus main protease	Maybridge 59 K	GOLD	2 hits of 50 tested	[159]
A1-antitrypsin aggregation	Combined commercial 1,200 K	ICM	10 hits of 68 tested	[160]

CDC25 phosphatase	Chembridge 413 K (313 K after filtering)	FRED/Surflex	99 hits of 1,500 tested	[161]
trans-Sialidase	Asinex GOLD and Platinum (300 K)	DOCK	5 hits of 32 tested	[162]
CDK2	975 K	RDOCK	38 of 1,121 tested	[163]
Protein phosphatase 2C	NCI diversity set (2 K)	Autodock	4 hits of ca. 100 tested	[164]
Protein arginine methyltransferase	Proprietary 9 K	GOLD		[165]
Protein arginine methyltransferase	NCI diversity set (2 K)	GOLD	7 hits of 30 tested	[166]
Arylalkylamine *N*-acetyltransferase	Commercial 1,200 K	GOLD	5 confirmed hits of 188 tested	[167]
SARS Cov 3C-like protease	Maybridge 59 K	Dock 4.0	23 hits of 93 tested	[168]
AHAS	NCI (164 K)	DOCK, Autodock	3 hits of 14 tested	[169]
Integrin avb3	SPECS 89 K	DOCK	14% hit rate of 50 tested	[170]

to differentiate between poses of the same molecule in a given receptor site. The top ranked pose, the pose with the best "docking score" might often actually be less similar to a crystallographically observed binding pose compared to a lower ranked pose and could be seen to be one with less binding energy when evaluated with a more accurate scoring function. This seemingly small error in choice of the best pose of a single molecule gets magnified and becomes substantial when the best docked poses of different molecules are compared, partly due to the breakdown in the comparability of the approximations in solvation and entropic terms across different ligands which can lead to incorrect rankings [171–174]. One could venture to say that a consensus SBVS view today would be that the inability to rank a database of ligands in order of their potential for binding to a given receptor site has more to do with our inability to score the binding affinity of a series of ligands in their predicted pose reliably than our ability to predict a reasonably accurate binding pose [175]. To that extent, one of the strategies proposed for effective SBVS is to generate a set of poses for a large collection of molecules rapidly using a well approximated but fast "docking function" and then rank with a more thorough but slower energy evaluation to rank molecules [176]. Extensive and continuing effort has focused on generating better scoring functions that better capture free energy differences between molecules, but can still operate fast enough to be of use to a high throughput docking experiment [175–178]. Results of head to head comparisons of docking and scoring using multiple docking scoring software frequently suggested that different scoring functions could be more effective for different receptors and this led to the drive towards consensus scoring functions [179–182].

Given that the focus of this chapter is SBVS and not a treatise on docking protocols, we give here a very brief and less than comprehensive coverage of docking algorithms and some of the commonly used docking software. For SBVS applications, the two most relevant pieces of information on the docking software would be the speed of the docking software and quality of pose(s) obtained. A number of packages are available, many of which have been applied to structure-based VS experiments and with success. Among the earliest attempted were incremental construction approaches, wherein the program attempts to exhaustively position the largest fragment in as many locations as possible with the active site, followed by adding subsequent fragments with suitable torsions. DOCK, FlexX, Hammerhead, and eHits are amongst the software that use this approach. Monte Carlo approaches to sampling the pose and conformation of the ligand are used by QXP, ICM, and PRODOCK and these tend to be slower. Evolutionary algorithms that improvise on preferred poses are used by docking software like GOLD, EP Dock and FITTER. GLIDE software uses a rough sampling initially and follows with a refined search using a more sophisticated scoring scheme. Surflex-Dock also performs a rough docking simulation to obtain seed poses which are refined further with a more rigorous scoring scheme. By using sequential docking simulations of varying rigor, the sampling approximately mimics the outcome of a more intensive search. Most of the docking software mentioned could be used to dock anywhere from a few hundred to 10,000 molecules in reasonable time and depending upon availability of processing power and parallelizability of the application, could screen up to 100,000 molecules within days. If the task is to dock millions of molecules, then it becomes faster to precompute a set of conformations for each molecule and limit docking to positioning the rigid ligand in the active site of the receptor. The FRED software from OpenEye used in conjunction with Omega, the conformer generator also available from OpenEye, takes this approach. Flexibase/FLOG also share the precomputed database approach.

With so much docking software to choose from, the SBVS practitioner is left with limited guidance in choice of docking and scoring options not to mention the critical postprocessing that has to bring the followed up hitlist to less than a hundred if the ligands or to be acquired through purchase for testing, and possibly a few thousand in a pharma setting. To that end, several studies have been published over the years that compare a subset, usually the most commonly available, docking and scoring applications, in a head-to-head comparison using datasets containing known hits and decoys for receptors where structural information is available and the enrichment could be studied carefully [106, 135, 173, 174, 176, 183–185]. If nothing else, these highlight the significant variation in performance of any package based on subtle variations in decisions about database construction, choice of data sets (both of active molecules and inactive decoys), program settings, and protein systems. In practice, most users rely on docking and scoring packages readily available to them, rather than try to find/use THAT ONE package that always works. The enrichment or the efficiency of the VS effort becomes more and more stringent as the proportion of compounds screened approaches 1% or less of the database of compounds screened. In these instances, for increasing the chances of success, one needs more than a protein structure, computing power, and software. Additional knowledge of the binding preferences of the active site in question can easily outstrip incremental advantages provided by one software over another and visual inspection and/or consensus approaches can aid in weeding out false positives effectively [21, 134].

The convenient shorthand in the community is that SBVS approaches are limited by "inadequate scoring functions". This is partly true, because in an SBVS experiment, due to the need for speed, the scoring functions do not perform a rigorous free energy calculation, and they will remain limited. Sampling, especially in situations where the ligand in question is increasingly complex with multiple rotatable bonds, can also be an issue. A significant other limitation is that generally only one protein structure is used, and held rigid. Movements of side chains or entire domains will not be modeled correctly and the extent an active site moves in response to a given ligand can be largely dependent on receptor mobility and ligand-dependent in addition to that. The resulting scores, whether fortuitously good or bad, will not be of the correct docked mode. In the latter case, even if appearing highly ranked in a hit list, this may be filtered out when inspected. Fortunately, there have been several attempts at addressing limited side chain mobility at least for situations of medium throughput and SBVS is becoming practical with inclusion of protein mobility [176, 186–190].

4.2 Case Studies

The study by Agrawal and coworkers [191] demonstrates key features in performing a structure-based VS. They searched for inhibitors of DNA primase from *E. coli*, an essential enzyme for bacterial reproduction with distant human homologs. Although a crystal structure of the DNA primase catalytic domain is available (1DDE) [192], this alone provides little insight into the most productive binding site. The GRID [193] software was used to identify three putative binding sites. The database to be searched was constructed from the catalogues of 20 vendors, and filtered to remove reactive functional groups, compounds with more than eight rotatable bonds, cLogP less than 5, and MW between 275 and 500. This resulted in a database of approximately 500,000 molecules. Representative 3D conformations were generated, protonated, and minimized. For each of the three sites identified, grids were generated with a 16-Å bounding box and a 20-Å enclosing box. Glide docking was performed, and the top 2,500 compounds as defined by the Glide SP score were inspected individually for feature complementarity, and correct ionization. A short list of 79 inhibitors was created, of which 68 were available for purchase. Of these, four inhibitors inhibited primase with an IC50 less than 100 μM.

A study by Alvesalo [194] and coworkers provides an interesting contrast in terms of methods and highlights some subtle considerations. In this case, they attempted to develop antimicrobial agents against *Chlamydia pneumoniae*. They chose to target dimethyladenosine transferase, but, because no crystal structure was available, chose to screen the structure of *Bacillus subtilis* RNA methyltransferase (1QAO) [195] as a surrogate. A database was constructed of molecules available from Specs and Maybridge, and contained 300,000 compounds after filtering for undesirable chemical groups. The database was docked into the protein binding site using FlexX [196], after which the top 2,000 molecules were inspected. From this set, 33 molecules were purchased. Of these, eight demonstrated >50% inhibition at 50 μM in a cell assay and represented two series of interest. This demonstrates that the use of a surrogate protein is viable if no exact crystal structure to the target of interest is available.

A VS study by Furci [197] and coworkers highlights the use of a third docking program, DOCK, against heme oxygenase from *Neisseria meningitides*, a Gram-negative pathogen. Heme oxygenase is an essential enzyme for heme utilization by the bacteria and blocking its function should arrest bacterial growth. The protein complex including heme (1P3T) [198] was subjected to molecular dynamics simulations with the heme removed to identify four suitable *apo* structures into which to dock the ligands. A database of 800,000 molecules was assembled from the supplier catalogues of Chembridge, Chemdiv, Maybridge, and SPECS. Compounds were docked into a single protein conformation to identify 50,000 molecules using the DOCK software. A second round of docking into all four representative protein conformations obtained from the molecular dynamics simulation allowed narrowing to the top 1,000 compounds (based on best docking score to any of the 4 protein conformations). This list was further narrowed by clustering and inspection, with 153 compounds being purchased for testing. Of the 153 compounds obtained, only 37 were soluble in DMSO or buffer, and of these, 10 interfered with the fluorescence polarization-based assay. Of the 27 tested, 8 exhibited inhibition of heme oxygenase with K_d values ranging from 12 to 240 μM. This study demonstrates the value of a sequential VS strategy, and also the way in which experimental considerations (i.e., compound solubility) can limit the overall impact of a VS study.

An example of a sequential docking strategy using different software is provided by a VS for CDC25 phosphatase inhibitors by Montes and coworkers [199]. CDC25 phosphatases play an important role in initiating cell cycle events; blockade may lead to useful anticancer effects. The structure of CDC25B (pdb code 1CWT) [200] was prepared for VS by adjusting the protonation states of various residues in the putative binding region. The 2005 release of the Chembridge database was filtered to remove compounds with undesirable reactive groups, leaving approximately 313,000 compounds. Up to 50 conformations per molecule were generated and the FRED software was used to dock the database into CDC25. The top 50,000 compounds were then redocked with full ligand flexibility using Surflex. The docked poses were then scored using either a receptor-specific Surflex function or with a receptor-specific function generated by LigScout. The top 450 molecules from each list, and the molecules that appeared in the top 3,000 molecules of both lists (total 1,500) were tested for enzyme inhibition. Of these, 99 showed at least 20% IC50 at 100 μM, with the most potent having an IC50 of 13 μM and showing inhibition in a HeLa cell assay. Overall, a number of interesting series were obtained, and the authors note the importance of consensus scoring in choosing their most active molecule.

5 Hybrid Workflows

As seen from the case studies described in the previous sections, many investigators use multiple complementary methods to reduce and refine their hit lists to manageable numbers. Often, an inspection step is included, which places a de facto upper bound to the size of the hitlists that are reviewed.

In this section, a number of case studies (Table 5) in which different types of VS methods are combined into a hybrid workflow. Often these combine a fast, ligand or pharmacophore-based method with a later docking method. The latter is useful at the inspection stage as it allows the molecule to be reviewed within the context of the protein binding site. A poor binding pose can be an indicator of a poor fit. Furthermore, possible interactions outside the scope of the molecules used to train the ligand-based method can be identified.

Table 5 Workflows that provide examples of the combined use of methods

Target	Database	Protocol	Outcome	Reference
Chorismate mutase	Proprietary 15 K	(a) UNITY, (b) FlexX	4 of 15 tested	[119]
ALK: anaplastic lymphoma kinase	Chembridge		'Numerous' hits of 2,677	[201]
HIV reverse transcriptase NNRTI	Derwent WDI, CAP complete, (67 K + 1,670 K)	(a) LigandScout, (b) Catalyst, c. Glide	5 of 6 tested	[202]
Cyclophilin A	ACD 2004 (296 K)	(a). ISISBASE, (b) FlexX, (c) Surflex	9 hits of 31 tested	[203]
DDAH	308 K	(a) rNN, (b) FlexX, (c) Sim searches	2 series of 109 purchased	[204]
Protein Kinase A, *Yersinia*	Proprietary (2,000 K)	(a) SVM kinase-like model, (b) FlexE	7 hits of 45 tested	[205]
CCR5	1,600 K	(a) pharmacophore (to 44 k), (b) GOLD, Surflex (homology model),	10 hits of 59 tested, 6 with functional response	[206]
HIV integrase	Chemnavigator (13,500 K)	(a) Catalyst (to 235 K), (b) Glide	9 hits of 88	[207]
MDM2/p53	NCI-3d (250 K filtered to 110 K)	(a)Web-pharmacophore (to 2.6 K), (b) GOLD	10 hits of 67 tested	[208]
ACE-2	3,800 K	(a) LigandScout/Catalyst, (b) ehits	6 hits of 17 tested	[209]
MGL/FAAH	Maybridge, Leadquest	(a) UNITY, (b) GOLD	No hits of 62 tested, 5 hits when retested in FAAH	[210]
HIV intergrase	Asinex GOLD (200 K)	(a) EIIP pharmacophore, (b) Catalyst, (c) Autodock	1 hit of 12 tested	[211]

5.1 Case Studies

An informative example of a hybrid workflow applied to HIV reverse transcriptase is provided by Barreca and coworkers [212]. Nonnucleoside reverse transcriptase inhibitors (NNRTI) bind to HIV reverse transcriptase and block viral replication. In this study, the Ligand Scout software was used to create a Catalyst pharmacophore from the protein complex of reverse transcriptase and Janssen R185545 [213] (1SUQ). This pharmacophore was used to search the World Drug Index (WDI, 67,000 molecules) and the Chemicals Available for Purchase (CAP, 1.7 million compounds). The molecules retrieved by the Catalyst pharmacophore with a fitness score greater than 3.0 included 521 from the WDI and 11,273 from the CAP. After filtering using Lipinski conditions, 9,345 remained. These were docked using Glide with SP scoring, and the best 1,000 hits were inspected individually. Interesting, novel compounds were evaluated for availability using the substructure capabilities in the Scifinder software, and six compounds were ordered and tested. Of these, five showed significant activity, with potency ranging from 0.2 to 4 μM.

A second study by Hartzoulakis et al. [204] also provides an example where the use of multiple methods facilitates an efficient search strategy. The target in this case was dimethylarginine dimethylaminohydrolase, an enzyme that modulates the nitric acid pathway in endothelial function, and may also control a cardiac risk factor. A bacterial ortholog from *Pseudomonas aeruginosa* may also contribute to pathogenicity in cystic fibrosis. A database of 308,000 commercial compounds was filtered to keep compounds with cLogP between −2 and 5, molecular weight less than 650, five or fewer hydrogen bond donors, ten or fewer acceptors, and ten or fewer rotatable bonds. This removed about 43,000 compounds. A reciprocal Near Neighbor clustering was used to select 35,000 compounds. These were docked into the active site of the DDAH enzyme from *Pseudomonas aeruginosa* [214] (1H70) using the FlexX software. The top 1,000 compounds were rescored using a combination of scoring methods, and the top 200 were inspected. Of the 109 selected, 90 were available and tested, of which three were interesting molecules, the most potent of which had an IC50 of 17 μM. This is an example in which clustering was used to reduce the number of compounds that were docked to a number consistent with the capacity of the FlexX program and their computing resources.

As an example of the utility of combining pharmacophore models and docking to select ligands from very large databases, the VS of Liao and coworkers [207] of HIV integrase offers an excellent example. In this case, the ChemNavigator database containing approximately 13.5 million compounds was searched. Thirty Catalyst pharmacophores were generated from known HIV integrase inhibitors, and all were used to search the database, resulting in about 235,000 hits. After filtering using Lipinski conditions and deduplication, the resulting 167,000 compounds were docked into a model of HIV integrase. The docked poses of the 1,500 top scoring compounds were inspected visually. After additional ADME models were applied and availability assessed, 88 compounds were obtained for testing.

Of these, eight compounds were assessed as active, with IC50 in their primary assay ranging from 37 to 780 μM.

6 Fragment-Based Virtual Screening

Many a pharmaceutical scientist would have at one time or another looked at a competitor's patent compound and looked for ways to find a lead that retains the activity of the competitor's compound but looks different enough not to infringe on the competitor's patent. A common strategy in such situations is to replace fragments in the molecule with isosteric fragments. These fragments could be small amounting to a few atoms or pieces that are over 100 Da or more in molecular weight. The FBVS discussion here is of fragments/substructures and does not pertain to fragments that are composed of five atoms or less. With increasing need in pharmaceutical research to have leads derived from more than one chemical class for a given target, to serve as a backup in case of unexpected failure of the lead candidate in the clinic which is attributable to compound class, researchers are sometimes looking to imitate their own compounds with a sufficiently different scaffold. FBVS is very similar to this strategy with a small twist.

FBVS presumes that all fragments of a tight binding ligand do not bind with the same ligand efficiency. While this is nothing new, in that computing properties of molecules using properties of their components is a very common occurrence in computational chemistry, fragment-based design successes in the recent literature [215–217] have given strong support to the notion that tight binding ligands can be obtained by starting from very ligand efficient albeit weak binding fragments and growing to larger ligands with high affinity when the added fragments are chosen with care so as not to compromise ligand efficiency significantly. When two fragments with affinity for a receptor are linked without restraining the ability of the fragments to bind to their respective preferred site on the receptor, the combined affinity is the sum of their binding affinities [218, 219].

For this to work, one has to have one or more seed ligands with at least moderate to high potency against the receptor. The more potent the seed ligand, the better. The molecule is then logically broken to fragments, typically at retrosynthetic bonds or if synthetic issues are not a key criteria, at rotatable bonds. Automated methods that take advantage of such fragmentation followed by piecewise similarity based retrieval followed by assembly have been reported [42, 220]. In cases where structural information is available for how the ligand binds to the target receptor, one could run energy computations to find the receptor affinity of the various fragments and weight the substructures and get better retrievals [221]. The rest of the VS is very straightforward. Two-dimensional similarity, pharmacophoric similarity or shape and electrostatic similarity could be used to find new fragments. The new fragments are linked together in an $n \times n$ matrix and tested for relevance by passing through a 2D similarity filter (to the seed molecule) or pharmacophore or protein–ligand interaction energy scoring filter (where structure is available, using docking and scoring) and other relevant filters. The resulting hits that look attractive enough could be synthesized or ordered for testing based on availability and synthesizability considerations and could also be used as idea generators.

The applicability of such VS in combination with tools available include situations where portions of any molecule need replacement with bioisosteric fragments. In this regard, BROOD software [105] and MOE [222] provide automated tools for fragment removal, replacement, and minimization to relieve any strain in the molecular assembly step and provide a database of fragments(isosteres) that could be enhanced in custom fashion by an enterprise as well. These software allow facile FBVS in 3D. Since this software has become available within the last 2 years, there seem to be a dearth of use cases in the published literature. However, anecdotal reports indicate that these are being used regularly in industry and the Websites of these two vendors provide adequate information for the inquisitive reader.

6.1 Case Study

Rummey et al. [223] searched replacements for the pyrrolidine present in their DPP-IV inhibitor searching a 10,000-molecule subset of small primary aliphatic amines extracted from the available chemical directory and visually inspected the top 500 of them. Four were selected for testing and two of them were novel hits.

Considering the power of these methods to retrieve novel molecules, it is only a matter of time before more successful reports are available.

7 Text-Mining as a Novel Virtual Screening Tool

"Can I use Google to find other molecules that have similar properties as my molecule?" could be an innocent question posed by someone new to computational chemistry. The irony of it is that all information about molecules are present in publications that are predominantly text, yet, the most powerful text-mining tool cannot retrieve it for us, at least at the present time, unless the molecular query is a simple name like glucose or pyrrolidine. To the present-day scientist, this might look something of an impossibility only if the person does not stop to think for a moment that the question would have hardly been comprehended by the average person only a decade ago. Text mining and natural language processing (NLP) a decade ago was not what it is today [224].

To a computer scientist, VS is nothing but another text mining, only the bits and bytes stored that contain molecular information adopt a format quite different from natural language and without adequate warning cannot be quickly interpreted. It is not that modern day text does not contain text that is not natural language, but that they are adequately flagged and do not stop the NLP software. For example, hyperlinks do not read like natural language but they are adequately flagged and are properly processed. In the case of chemical structures, the material to be searched, the algorithms, used and retrieval techniques are geared towards structure perception and manipulation although the information is still stored and operated on as bits and bytes. This limitation exists because molecular information is not expressed in natural language in an easily perceptible form, and where we do express them, in patents for example, it is so convoluted that very few people attempt to read and decipher the chemical structure or composition by reading the IUPAC name detailed in a patent. Everyone reaches for a translator, nowadays inevitably the appropriate software, that could translate the name into the familiar chemical structure form. Unfortunately the one line smiles representation of a molecule did not come into vogue soon enough and computing facilities did not exist to encourage the broad range of scientists to represent every structure to be associated with its smiles in written documents with appropriate flags to enable software to interpret it correctly.

7.1 *Current Limitations*

One of the greatest limitations of searching for molecules is the fact that the database is finite. Several forms of text similarity are a part of the strategies used by people not trained in science and those easy similarity search strategies are not available to the scientist searching through molecules. Unless the database is prepared in a specific format and made available, searching cannot proceed. Search results are curation dependent and associations are limited by curation capabilities and subject to errors and biases introduced at the point of curation [224]. To give a simple example, if the curator errs and associates a wrong number with a molecule structure in the main database, regardless of how many other documents carry the correct information, people will repeatedly extract the wrong information because the association cannot be deciphered using NLP from other corporate documents. The rate of publishing is exploding, and curation is limiting. Imagine entering the smiles string for a molecular fragment in Google and get 300 references all discussing various pieces of information about it! Imagine replacing one of the carbon with an asterisk and seeing many analogs and information about them as well.

7.2 *The Rewards of Storing Molecular Structures in NLP Searchable Form*

Screening brings back a rich variety of information, not just what the curator put in the database. Suddenly a chemist can read everything about a molecule ever printed, not just what someone decided to associate it with. Distant associations – A related to B and B related to C might mean A related to C– will become apparent.

Chemical structural information is one of the missing pieces in the great effort to bring biomedical research into the realm of twenty-first century information extraction and knowledge discovery paradigms. Proteins, genes, diseases, and chemical compounds constitute the major entities extracted in the biomedical domain. The ability to read structure information and substructure information and their association to other entities could have a major impact on toxicity information in particular and ADMET data in general.

7.3 Potential Long Term Solutions

How do we do it? Every 2D structure reference created in the future should have a hyperlink to a canonical smiles string. Smiles readers should be freeware so when mousing over the molecule reference, the structure pops up. Start representing structures today and 15 years from now, our 2D VS efforts will look very different. The main added advantage will be that the data associated with every structure will be available for natural language processing software from which to process and extract information. Structures themselves can be searched in unforeseen ways. This will bring information about molecules in an unprecedented fashion to the average reader.

7.4 Potential Short Term Solutions

There are few short term solutions that we can think of. The technology for accessing publications underwent a dramatic makeover in the last decade, moving from predominantly paper to predominantly electronic through a coordinated set of efforts from publishers and consumers (in this case scientific research users) alike. A study of how such a transition was successfully handled could provide clues on how to make it happen.

8 Summary

VS continues to be a growing area, fueled by the dramatic increase in affordably priced computing capability, and the development of better algorithms and software. Its position as a cost-effective alternative to high-throughput screening, the traditional engine for lead identification for pharmaceutical discovery, is bound to rise, despite the technology advancements in screening through ultrahigh throughput methods, miniaturization, and automation. This is partly due to the high cost of personnel and reagents, both of which are needed in larger supply for HTS compared to VS. However, as the field stands today, one would be very justified in stating that VS is nowhere near replacing experimental screening methods and this is mostly due to the inconsistency of success in finding leads using VS. Many factors, the target itself and the information available to prime the VS effort being the major ones, the technique, the software and the expertise of the screener being the minor ones, influence the success rate. This continues to be a fast growing field, and the recent trends and progress in identifying the major challenges and addressing them effectively both at the scientific and algorithmic levels bodes well for the future of this method.

8.1 Virtual Screening Strategy

There has been considerable debate within the community and in the literature about the relative merits of ligand-based vs protein structure-based screening. In principle, the protein-based screen should provide the broadest access to novel chemotypes that could interact with the relevant binding site. The 2D ligand-based methods are often best at retrieving hits chemically similar (same or highly related scaffold, comparable pendant groups) to the query molecules. There have been efforts to develop measures of chemical similarity based on 2D graphs alone that better generalize the hits retrieved to compounds that include dissimilar but acceptable alternative scaffolds. However, these approaches tend to retrieve a large number of false positives; setting a similarity threshold to include these more dissimilar-but-acceptable hits often leads to the inclusion of far more dissimilar-but-unacceptable hits, leading to less enrichment. A third option has been the emergence of 3D similarity methods. These appear to provide a compromise leading to a balanced retrieval of both analogues and compounds containing alternative chemical scaffolds [1, 106].

Optimal strategy rests in balancing a mix of techniques and shaping the workflow for a given VS based on the information available, the perceived strengths and limitations of various techniques, and the time and effort needed. Clearly, in the absence of a protein crystal structure or acceptable homology model, ligand-based screening is the obvious option. At the other extreme, in the absence of known ligands, a protein-based screen could be contemplated. When one or more protein crystal structures are available, as well as a number of ligands that have been identified either from the literature or by some previous experimental effort, a priori, the all-out approach would be to bring to bear all available techniques to the problem. However, ligand-based screening often requires less preparation and less analysis of results, thus being sparing of the computational chemist's time and first one to get results out. Protein-based screening generally requires more time to prepare and validate the simulation, and to analyze the results, often including visual inspection to ensure that docked modes are acceptable. The choice of strategy then requires a balance between the enrichment that is expected, the anticipated novelty of the hits, and the time and effort available to invest in the effort.

As general guidance, we would suggest the following guidelines:

1.

Include VS in a lead discovery strategy whenever possible. Computational VS is low cost. It is typically performed by a single scientist who employs multiple processors, typically LINUX clusters now available at commodity prices. Of the many resources needed in the drug discovery process, processor time belongs in the inexpensive category. VS also brings considerable benefit. Many of the methods available offer some enrichment over purely random screening, and often offer significant enrichment.

2. *Test a substantial number of compounds*. VS methods generally offer enrichment, but most ranked hit lists contain a significant proportion of false positives. Hitlists should be scaled to 1–5% of the compounds in the virtual library screened. In many real world situations, the computational chemist is being asked to choose lists of compounds representing 0.1% or less of the compounds screened (e.g., the "best 100" of 100,000 compounds). Typically, VS methods have been validated considering 1%, 5%, or 10% of the total number of compounds in the VS collection. By following up on more compounds, one increases the probability of impact from VS.
3. *Include a 3D ligand-based method*. In our internal efforts across two companies, we arrived at the same conclusion as the Merck researchers [106] that a 3D similarity method appears to offer a good balance between effort expended and the number and novelty of hits generated.
4. *Automate*. Much of the human effort in VS arises at the point of combining various hit lists, followed by scoring and selection. The more this can be automated, the more efficient the VS experiment becomes.
5. *Integrate*. An effective strategy is to view VS as an approach to identifying chemical matter that is complementary to wet methods. This opens up potential symbiosis between the VS benefiting from the HTS, or alternatively, HTS benefiting from early hits identified by VS. Such a complementary view cannot be overemphasized given that the role of VS in drug discovery is often looked upon as competitive with high throughput screening or focused subset screening. However, the lower cost and faster completion times should make VS acceptable even with lower enrichment numbers. The savings in cost and time to obtain a hitlist of active compounds can be significant if additional factors like the cost and time of adaptation of an assay for HTS purposes, compound depletion in the collection due to HTS, level of false positives from HTS created by mechanical and measurement errors are considered.
6. *Whenever possible, inspect the hitlist*. Within the literature, there is a surprising number of instances in which small numbers of compounds were ultimately ordered. This inevitably requires individual inspection of compounds. In this situation, applying all relevant simulations and any hypotheses based on prior knowledge about key features are key contributors to higher enrichment. Where it is possible to order a larger VS hitlist for testing, some additional tolerance in favor of serendipity is beneficial. (For example, lowering the VDW radius of ligands or proteins to allow for possible protein motion or just ignoring small steric clashes.)

References

1. Hawkins PCD, Skillman AG, Nicholls A (2007) J Med Chem 50:74
2. Hert J, Willett P, Wilton DJ, Acklin P, Azzaoui K, Jacoby E, Schuffenhauer A (2005) J Med Chem 48:7049
3. Xie X-Q, Chen J-Z (2008) J Chem Inf Model 48:465
4. Perekhodtsev GD (2007) QSAR Comb Sci 26:346
5. Willett P (2006) Drug Disc Today 11:1046
6. Eckert H, Bajorath J (2007) Drug Disc Today 12:225
7. Cramer RD, Jilek RJ, Guessregen S, Clark SJ, Wendt B, Clark RD (2004) J Med Chem 47:6777
8. Cramer RD (2006) Expert Opin Drug Disc 1:311
9. Jalaie M, Shanmugasundaram V (2006) Mini-Rev Med Chem 6:1159
10. Reddy AS, Pati SP, Kumar PP, Pradeep HN, Sastry GN (2007) Curr Protein Pept Sci 8:329
11. Hert J, Willett P, Wilton DJ (2006) Chemogenomics 133
12. Klebe G (2006) Drug Disc Today 11:580
13. Triballeau N, Acher F, Brabet I, Pin J-P, Bertrand H-O (2005) J Med Chem 48:2534
14. Bayly CI, Truchon J-F (2007) Abstracts of Papers, 234th ACS National Meeting, Boston, MA, United States, August 19–23
15. Wikipedia www.wikipedia.org
16. Bewick V, Cheek L, Ball J (2004) Critical Care 8:508
17. Bewick V, Cheek L, Ball J (2004) http://www.pubmedcentral.nih.gov/articlerender.fcgi?artid=1065080
18. Seifert MHJ (2008) J Chem Inf Model 48:602
19. Irwin JJ, Shoichet BK (2005) J Chem Inf Comput Sci 45:177
20. Huang N, Shoichet BK, Irwin JJ (2006) J Med Chem 49:6789
21. Perola E (2006) Proteins Struct Funct Bioinf 64:422
22. Al-Gharabli SI, Shah STA, Weik S, Schmidt MF, Mesters JR, Kuhn D, Klebe G, Hilgenfeld R, Rademann J (2006) ChemBioChem 7:1048
23. Alvesalo JKO, Siiskonen A, Vainio MJ, Tammela PSM, Vuorela PM (2006) J Med Chem 49:2353
24. Boehm M, Wu T-Y, Claussen H, Lemmen C (2008) J Med Chem 51:2468
25. Gasteiger J, Rudolph C, Sadowski J (1990) Tetrahedron Comput Methodol 3:537
26. Pearlman RS (1993) 3D QSAR Drug Des 41
27. OpenEyeScientificSoftware (2006). http://www.eyesopen.com/
28. AccelrysInc (2008) http://www.accelrys.com
29. Lipinski CA, Lombardo F, Dominy BW, Feeney PJ (1997) Adv Drug Deliv Rev 23:3
30. Wenlock MC, Austin RP, Barton P, Davis AM, Leeson PD (2003) J Med Chem 46:1250
31. Lipinski CA (2004) Drug Disc Today Technol 1:337
32. Veber DF, Johnson SR, Cheng H-Y, Smith BR, Ward KW, Kopple KD (2002) J Med Chem 45:2615
33. Hann MM, Leach AR, Harper G (2001) J Chem Inf Comput Sci 41:856
34. Oprea TI, Davis AM, Teague SJ, Leeson PD (2001) J Chem Inf Comput Sci 41:1308
35. Vainio Mikko J, Johnson Mark S (2007) J Chem Inf Model 47:2462
36. Jorgensen WL (2004) Science 303:1813
37. Davis AM, Keeling DJ, Steele J, Tomkinson NP, Tinker AC (2005) Curr Top Med Chem 5:421
38. Chen B, Harrison RF, Papadatos G, Willett P, Wood DJ, Lewell XQ, Greenidge P, Stiefl N (2007) J Comput Aided Mol Des 21:53
39. Yeap SK, Snarey M, Federico C (2003) Abstracts of Papers, 225th ACS National Meeting, New Orleans, LA, United States, March 23–27
40. Bender A, Glen RC (2005) J Chem Inf Model 45:1369

41. Kogej T, Engkvist O, Blomberg N, Muresan S (2006) J Chem Inf Model 46:1201
42. Joergensen AMM, Langgaard M, Gundertofte K, Pedersen JT (2006) QSAR Comb Sci 25:221
43. Muchmore SW, Debe DA, Metz JT, Brown SP, Martin YC, Hajduk PJ (2008) J Chem Inf Model 48:941
44. Fechner U, Franke L, Renner S, Schneider P, Schneider G (2004) J Comput Aided Mol Des 17:687
45. Weininger D (1994) Spec Publ R Soc Chem 142:67
46. Shemetulskis NE, Weininger D, Blankley CJ, Yang JJ, Humblet C (1996) J Chem Inf Comput Sci 36:862
47. Young SS, Gombar VK, Emptage MR, Cariello NF, Lambert C (2002) Chemom Intell Lab Syst 60:5
48. Hert J, Willett P, Wilton DJ, Acklin P, Azzaoui K, Jacoby E, Schuffenhauer A (2004) Org Biomol Chem 2:3256
49. PipeLinePilot (2008) http://accelrys.com/products/scitegic/
50. Sun H (2005) J Med Chem 48:4031
51. Chemical Computing Group M, Quebec, Canada (2005) www.chemcomp.com
52. Ewing T, Baber JC, Feher M (2006) J Chem Inf Model 46:2423
53. Bender A, Mussa HY, Gill GS, Glen RC (2004) J Med Chem 47:6569
54. Bender A, Jenkins JL, Glick M, Deng Z, Nettles JH, Davies JW (2006) J Chem Inf Model 46:2445
55. Zimmermann M, Hindle SA, Naumann T, Matter H, Hessler G, Baringhaus K-H, Lemmen C, Gastreich M, Rarey M (2003) Abstracts of Papers, 226th ACS National Meeting, New York, NY, United States, September 7–11
56. Hessler G, Zimmermann M, Matter H, Evers A, Naumann T, Lengauer T, Rarey M (2005) J Med Chem 48:6575
57. Blower PE Jr, Cross K, Fligner M, Verducci J (2002) Abstracts of Papers, 223rd ACS National Meeting, Orlando, FL, United States, April 7–11
58. Digital_Chemistry http://www.digitalchemistry.co.uk/
59. Roberts G, Myatt GJ, Johnson WP, Cross KP, Blower PE Jr (2000) J Chem Inf Comput Sci 40:1302
60. Pearlman RS, Smith KM (1998) Perspect Drug Disc Des 9/11:339
61. Pearlman RS, Smith KM (1999) J Chem Inf Comput Sci 39:28
62. Beno BR, Mason JS (2001) Drug Disc Today 6:251
63. Pirard B, Pickett SD (2000) J Chem Inf Comput Sci 40:1431
64. Shanmugasundaram V, Maggiora GM, Lajiness MS (2005) J Med Chem 48:240
65. Boecker A, Sasse BC, Nietert M, Stark H, Schneider G (2007) ChemMedChem 2:1000
66. Carosati E, Mannhold R, Wahl P, Hansen JB, Fremming T, Zamora I, Cianchetta G, Baroni M (2007) J Med Chem 50:2117
67. Carosati E, Budriesi R, Ioan P, Ugenti MP, Frosini M, Fusi F, Corda G, Cosimelli B, Spinelli D, Chiarini A, Cruciani G (2008) J Med Chem 51:5552
68. Franke L, Schwarz O, Mueller-Kuhrt L, Hoernig C, Fischer L, George S, Tanrikulu Y, Schneider P, Werz O, Steinhilber D, Schneider G (2007) J Med Chem 50:2640
69. Shoda M, Harada T, Yano K, Stahura FL, Himeno T, Shiojiri S, Kogami Y, Kouji H, Bajorath J (2007) ChemMedChem 2:515
70. Stanton DT, Ankenbauer J, Rothgeb D, Draper M, Paula S (2007) Bioorg Med Chem 15:6062
71. Yamazaki K, Kusunose N, Fujita K, Sato H, Asano S, Dan A, Kanaoka M (2006) Bioorg Med Chem Lett 16:1371
72. Manetti F, Magnani M, Castagnolo D, Passalacqua L, Botta M, Corelli F, Saddi M, Deidda D, De Logu A (2006) ChemMedChem 1:973
73. Franke L, Schwarz O, Mueller-Kuhrt L, Hoernig C, Fischer L, George S, Tanrikulu Y, Schneider P, Werz O, Steinhilber D, Schneider G (2007) J Med Chem 50:2640

74. AnalytiCon Drug Discovery http://www.ac-discovery.com
75. Schneider G, Neidhart W, Giller T, Schmid G (1999) Angew Chem Int Ed 38:2894
76. Fechner U, Franke L, Renner S, Schneider P, Schneider G (2003) J Comput Aided Mol Des 17:687
77. Fechner U, Franke L, Renner S, Schneider P, Schneider G (2003) J Comput Aided Mol Des 17:687
78. Mannhold R, Berellini G, Carosati E, Benedetti P (2005) In: Cruciani G (ed) Molecular interaction fields in drug discovery (Methods and Principles in Medicinal Chemistry), vol 27. Wiley, Weinheim, Germany, p 173
79. Pastor M, Cruciani G, McLay I, Pickett SD, Clementi S (2000) J Med Chem 43:3233
80. Baroni M, Cruciani G, Sciabola S, Perruccio F, Mason JS (2007) J Chem Inf Model 47:279
81. Van Drie JH (2007) Internet Electron J Mol Des 6:271
82. Kubinyi H (1999) J Recept Signal Transduct Res 19:15
83. Kubinyi H (2003) Nat Rev Drug Disc 2:665
84. Van Drie J (2006) Abstracts of Papers, 231st ACS National Meeting, Atlanta, GA, United States, March 26–30, 2006
85. van Drie JH (2005) Drug Disc Ser 1:157
86. Martin YC (2006) Compr Med Chem II 4:515
87. Martin YC, Bures MG, Danaher EA, DeLazzer J, Kim KH, Lico I, Pavlik PA (1993) Comput Aided Innovation New Mater 2, Proc Int Conf Exhib Comput Appl Mater Mol Sci Eng 1117
88. Guner OF (2002) Curr Top Med Chem 2:1321
89. Wermuth CG (2006) Methods Princ Med Chem 32:3
90. Cohen C, Fischel O, Cohen E (2006) Chem Biol Drug Des 67:182
91. Kurogi Y, Guner OF (2001) Curr Med Chem 8:1035
92. Leitao A, Andricopulo AD, Montanari CA (2007) Curr Methods Med Chem Biol Phys 1:61
93. Cramer RD, Wendt B (2007) J Comput Aided Mol Des 21:23
94. Jilek R, Cramer RD (2006) Abstracts of Papers, 232nd ACS National Meeting, San Francisco, CA, United States, Sept. 10–14
95. Polanski J (2006) Expert Opin Drug Disc 1:693
96. Dixon SL, Smondyrev AM, Rao SN (2006) Chem Biol Drug Des 67:370
97. Oprea TI (2004) Comput Med Chem Drug Disc 571
98. Akamatsu M (2002) Curr Top Med Chem 2:1381
99. Kubinyi H (2008) Comput Struct Approaches Drug Disc 24
100. Debnath AK (2001) Mini-Rev Med Chem 1:187
101. Kubinyi H (1997) Drug Disc Today 2:457
102. Kubinyi H (1997) Drug Disc Today 2:538
103. Connolly Martin Y (1998) Perspect Drug Disc Des 12/14:3
104. Jain AN (2000) J Comput Aided Mol Des 14:199
105. Openeye_Scientific_Software http://en.wikipedia.org/wiki/OpenEye_Scientific_Software
106. McGaughey GB, Sheridan RP, Bayly CI, Culberson JC, Kreatsoulas C, Lindsley S, Maiorov V, Truchon J-F, Cornell WD (2007) J Chem Inf Model 47:1504
107. Wolber G, Langer T (2005) J Chem Inf Comput Sci 45:160
108. Cheeseright T, Mackey M, Rose S, Vinter A (2007) Expert Opin Drug Disc 2:131
109. Cheeseright T, Mackey M, Rose S, Vinter A (2006) J Chem Inf Model 46:665
110. Schuster D, Maurer EM, Laggner C, Nashev LG, Wilckens T, Langer T, Odermatt A (2006) J Med Chem 49:3454
111. Purushottamachar P, Khandelwal A, Chopra P, Maheshwari N, Gediya LK, Vasaitis TS, Bruno RD, Clement OO, Njar VCO (2007) Bioorg Med Chem 15:3413
112. Larbig G, Pickhardt M, Lloyd DG, Schmidt B, Mandelkow E (2007) Curr Alzheimer Res 4:315
113. Michaux C, de Leval X, Julemont F, Dogne J-M, Pirotte B, Durant F (2006) Eur J Med Chem 41:1446
114. Bhattacharjee AK (2006) Lett Drug Des Disc 3:219

115. Bohacek R, Boosalis MS, McMartin C, Faller DV, Perrine SP (2006) Chem Biol Drug Des 67:318
116. Jensen AA, Begum N, Vogensen SB, Knapp KM, Gundertofte K, Dzyuba SV, Ishii H, Nakanishi K, Kristiansen U, Stromgaard K (2007) J Med Chem 50:1610
117. Ray NC, Clark RD, Clark DE, Williams K, Hickin HG, Crackett PH, Dyke HJ, Lockey PM, Wong M, Devos R, White A, Belanoff JK (2007) Bioorg Med Chem Lett 17:4901
118. Bhattacharjee AK, Nichols DA, Gerena L, Roncal N, Gutteridge CE (2007) Med Chem 3:317
119. Agrawal H, Kumar A, Bal NC, Siddiqi MI, Arora A (2007) Bioorg Med Chem Lett 17:3053
120. Lu IL, Huang C-F, Peng Y-H, Lin Y-T, Hsieh H-P, Chen C-T, Lien T-W, Lee H-J, Mahindroo N, Prakash E, Yueh A, Chen H-Y, Goparaju CMV, Chen X, Liao C-C, Chao Y-S, Hsu JTA, Wu S-Y (2006) J Med Chem 49:2703
121. Bhattacharjee AK (2007) Expert Opin Drug Disc 2:1115
122. Tervo AJ, Suuronen T, Kyrylenko S, Kuusisto E, Kiviranta PH, Salminen A, Leppaenen J, Poso A (2006) J Med Chem 49:7239
123. Doddareddy MR, Choo H, Cho YS, Rhim H, Koh HY, Lee J-H, Jeong S-W, Pae AN (2007) Bioorg Med Chem 15:1091
124. Lemmen C, Lengauer T, Klebe G (1998) J Med Chem 41:4502
125. Cruciani G, Pastor M, Guba W (2000) Eur J Pharm Sci 11:29
126. Norin M, Sundstrom M (2002) Trends Biotechnol 20:79
127. Chayen NE (2004) Curr Opin Struct Biol 14:577
128. Chayen NE (2002) Trends Biotechnol 20:98
129. DesJarlais RL, Sheridan RP, Dixon JS, Kuntz ID, Venkataraghavan R (1986) J Med Chem 29:2149
130. DesJarlais RL, Cummings MD, Gibbs AC (2007) Front Drug Des Disc 3:81
131. Kubinyi H, Boehm HJ (1997) Book of Abstracts, 214th ACS National Meeting, Las Vegas, NV, September 7–11 CINF
132. Waszkowycz B (2008) Drug Disc Today 13:219
133. Polgar T, Keseru GM (2007) Front Drug Des Disc 3:477
134. Brenk R, Klebe G (2006) Methods Princ Med Chem 32:171
135. Perola E, Walters WP, Charifson PS (2004) Proteins Struct Funct Bioinf 56:235
136. Cross JB, Jalaie M, Gantt SL, Joshi NA, Wild DJ, Snow ME, Narasimhan LS (2004) Abstracts of Papers, 227th ACS National Meeting, Anaheim, CA, United States, March 28–April 1, 2004 COMP
137. Lemmen C, Rarey M, Schellhammer I (2007) (Biosolveit GmbH, Germany) Application: WO 2007071411, p 27
138. Schellhammer I, Rarey M (2004) Proteins Struct Funct Bioinf 57:504
139. Jansen JM, Martin EJ (2004) Curr Opin Chem Biol 8:359
140. Smith R, Hubbard RE, Gschwend DA, Leach AR, Good AC (2003) J Mol Graphics Modell 22:41
141. Good AC, Cheney DL, Sitkoff DF, Tokarski JS, Stouch TR, Bassolino DA, Krystek SR, Li Y, Mason JS, Perkins TDJ (2003) J Mol Graphics Modell 22:31
142. Good AC, Cheney DL (2003) J Mol Graphics Modell 22:23
143. Agarwal A, Louise-May S, Thanassi JA, Podos SD, Cheng J, Thoma C, Liu C, Wiles JA, Nelson DM, Phadke AS, Bradbury BJ, Deshpande MS, Pucci MJ (2007) Bioorg Med Chem Lett 17:2807
144. Bakir F, Kher S, Pannala M, Wilson N, Nguyen T, Sircar I, Takedomi K, Fukushima C, Zapf J, Xu K, Zhang S-H, Liu J, Morera L, Schneider L, Sakurai N, Jack R, Cheng J-F (2007) Bioorg Med Chem Lett 17:3473
145. Brooks WH, McCloskey DE, Daniel KG, Ealick SE, Secrist JA III, Waud WR, Pegg AE, Guida WC (2007) J Chem Inf Model 47:1897
146. Cherkasov A, Ban F, Li Y, Fallahi M, Hammond GL (2006) J Med Chem 49:7466

147. de Graaf C, Oostenbrink C, Keizers PHJ, van der Wijst T, Jongejan A, Vermeulen NPE (2006) J Med Chem 49:2417
148. Desai PV, Patny A, Gut J, Rosenthal PJ, Tekwani B, Srivastava A, Avery M (2006) J Med Chem 49:1576
149. Dooley AJ, Shindo N, Taggart B, Park J-G, Pang Y-P (2006) Bioorg Med Chem Lett 16:830
150. Foloppe N, Fisher LM, Howes R, Potter A, Robertson AGS, Surgenor AE (2006) Bioorg Med Chem 14:4792
151. Furci LM, Lopes P, Eakanunkul S, Zhong S, MacKerell AD Jr, Wilks A (2007) J Med Chem 50:3804
152. Hirayama K, Aoki S, Nishikawa K, Matsumoto T, Wada K (2007) Bioorg Med Chem 15:6810
153. Kenyon V, Chorny I, Carvajal WJ, Holman TR, Jacobson MP (2006) J Med Chem 49:1356
154. Knox AJS, Meegan MJ, Sobolev V, Frost D, Zisterer DM, Williams DC, Lloyd DG (2007) J Med Chem 50:5301
155. Leban J, Baierl M, Mies J, Trentinaglia V, Rath S, Kronthaler K, Wolf K, Gotschlich A, Seifert MHJ (2007) Bioorg Med Chem Lett 17:5858
156. Li J, Chen J, Gui C, Zhang L, Qin Y, Xu Q, Zhang J, Liu H, Shen X, Jiang H (2006) Bioorg Med Chem 14:2209
157. Liu H, Gao Z-B, Yao Z, Zheng S, Li Y, Zhu W, Tan X, Luo X, Shen J, Chen K, Hu G-Y, Jiang H (2007) J Med Chem 50:83
158. Louise-May S, Yang W, Nie X, Liu D, Deshpande MS, Phadke AS, Huang M, Agarwal A (2007) Bioorg Med Chem Lett 17:3905
159. Lu IL, Mahindroo N, Liang P-H, Peng Y-H, Kuo C-J, Tsai K-C, Hsieh H-P, Chao Y-S, Wu S-Y (2006) J Med Chem 49:5154
160. Mallya M, Phillips RL, Saldanha SA, Gooptu B, Brown SCL, Termine DJ, Shirvani AM, Wu Y, Sifers RN, Abagyan R, Lomas DA (2007) J Med Chem 50:5357
161. Montes M, Braud E, Miteva MA, Goddard M-L, Mondesert O, Kolb S, Brun M-P, Ducommun B, Garbay C, Villoutreix BO (2008) J Chem Inf Model 48:157
162. Neres J, Bonnet P, Edwards PN, Kotian PL, Buschiazzo A, Alzari PM, Bryce RA, Douglas KT (2007) Bioorg Med Chem 15:2106
163. Richardson CM, Nunns CL, Williamson DS, Parratt MJ, Dokurno P, Howes R, Borgognoni J, Drysdale MJ, Finch H, Hubbard RE, Jackson PS, Kierstan P, Lentzen G, Moore JD, Murray JB, Simmonite H, Surgenor AE, Torrance CJ (2007) Bioorg Med Chem Lett 17:3880
164. Rogers JP, Beuscher AE, Flajolet M, McAvoy T, Nairn AC, Olson AJ, Greengard P (2006) J Med Chem 49:1658
165. Spannhoff A, Machmur R, Heinke R, Trojer P, Bauer I, Brosch G, Schuele R, Hanefeld W, Sippl W, Jung M (2007) Bioorg Med Chem Lett 17:4150
166. Spannhoff A, Heinke R, Bauer I, Trojer P, Metzger E, Gust R, Schuele R, Brosch G, Sippl W, Jung M (2007) J Med Chem 50:2319
167. Szewczuk LM, Saldanha SA, Ganguly S, Bowers EM, Javoroncov M, Karanam B, Culhane JC, Holbert MA, Klein DC, Abagyan R, Cole PA (2007) J Med Chem 50:5330
168. Tsai K-C, Chen S-Y, Liang P-H, Lu IL, Mahindroo N, Hsieh H-P, Chao Y-S, Liu L, Liu D, Lien W, Lin T-H, Wu S-Y (2006) J Med Chem 49:3485
169. Wang JG, Xiao YJ, Li YH, Liu XH, Li ZM (2006) Chin Chem Lett 17:1555
170. Zhou Y, Peng H, Ji Q, Qi J, Zhu Z, Yang C (2006) Bioorg Med Chem Lett 16:5878
171. Kuhn B, Gerber P, Schulz-Gasch T, Stahl M (2005) J Med Chem 48:4040
172. Schulz-Gasch T, Stahl M (2004) Drug Disc Today Technol 1:231
173. Schulz-Gasch T, Stahl M (2003) J Mol Model 9:47
174. Stahl M, Rarey M (2001) J Med Chem 44:1035
175. Spyrakis F, Kellogg GE, Amadasi A, Cozzini P (2007) Front Drug Des Disc 3:317
176. Zhou Z, Felts AK, Friesner RA, Levy RM (2007) J Chem Inf Model 47:1599
177. Spyrakis F, Amadasi A, Fornabaio M, Abraham DJ, Mozzarelli A, Kellogg GE, Cozzini P (2007) Eur J Med Chem 42:921

178. Montes M, Miteva MA, Villoutreix BO (2007) Proteins Struct Funct Bioinf 68:712
179. Teramoto R, Fukunishi H (2007) J Chem Inf Model 47:526
180. Betzi S, Suhre K, Chetrit B, Guerlesquin F, Morelli X (2006) J Chem Inf Model 46:1704
181. Feher M (2006) Drug Disc Today 11:421
182. Baber JC, Shirley WA, Gao Y, Feher M (2006) J Chem Inf Model 46:277
183. Joseph-McCarthy D, Baber JC, Feyfant E, Thompson DC, Humblet C (2007) Curr Opin Drug Disc Dev 10:264
184. Onodera K, Satou K, Hirota H (2007) J Chem Inf Model 47:1609
185. Kellenberger E, Rodrigo J, Muller P, Rognan D (2004) Proteins Struct Funct Bioinf 57:225
186. Bowman AL, Lerner MG, Carlson HA (2007) J Am Chem Soc 129:3634
187. Lerner MG, Bowman AL, Carlson HA (2007) J Chem Inf Model 47:2358
188. Koska J, Spassov VZ, Maynard AJ, Yan L, Austin N, Flook PK, Venkatachalam CM (2008) J Chem Inf Model 48:1965
189. Venkatachalam CM (2007) Abstracts, 41st Western Regional Meeting of the American Chemical Society, San Diego, CA, United States, October 9–13 GEN
190. Cavasotto CN, Orry AJW (2007) Curr Top Med Chem 7:1006
191. Agrawal H, Kumar A, Bal NC, Siddiqi MI, Arora A (2007) Bioorg Med Chem Lett 17:3053
192. Keck JL, Roche DD, Lynch AS, Berger JM (2000) Science 287:2482
193. Goodford P (2006) In: Cruciani G, Mannhold R, Kubinyi H, Folkers G (eds) Molecular interaction fields: applications in drug discovery and ADME prediction. Wiley, New York, p 3
194. Alvesalo JKO, Siiskonen A, Vainio MJ, Tammela PSM, Vuorela PM (2006) J Med Chem 49:2353
195. Schluckebier G, Zhong P, Stewart KD, Kavanaugh TJ, Abad-Zapatero C (1999) J Mol Biol 2:277
196. Rarey M, Kramer B, Lengauer T, Klebe G (1996) J Mol Biol 3:470
197. Furci LM, Lopes P, Eakanunkul S, Zhong S, MacKerell AD Jr, Wilks A (2007) J Med Chem 50:3804
198. Friedman J, Lad L, Deshmukh R, Li H, Wilks A, Poulos TL (2003) J Biol Chem 278:34654
199. Montes M, Braud E, Miteva MA, Goddard M-L, Mondésert O, Kolb S, Brun M-P, Ducommun B, Garbay C, Villoutreix BO (2008) J Chem Inf Model 48:157
200. Reynolds RA, Yem AW, Wolfe CL, Deibel MR Jr, Chidester CG, Watenpaugh KD (1999) J Mol Biol 293:559
201. Li R, Xue L, Zhu T, Jiang Q, Cui X, Yan Z, McGee D, Wang J, Gantla VR, Pickens JC, McGrath D, Chucholowski A, Morris SW, Webb TR (2006) J Med Chem 49:1006
202. Barreiro G, Guimaraes CRW, Tubert-Brohman I, Lyons TM, Tirado-Rives J, Jorgensen WL (2007) J Chem Inf Model 47:2416
203. Guichou J-F, Viaud J, Mettling C, Subra G, Lin Y-L, Chavanieu A (2006) J Med Chem 49:900
204. Hartzoulakis B, Rossiter S, Gill H, O'Hara B, Steinke E, Gane PJ, Hurtado-Guerrero R, Leiper JM, Vallance P, Rust JM, Selwood DL (2007) Bioorg Med Chem Lett 17:3953
205. Hu X, Prehna G, Stebbins CE (2007) J Med Chem 50:3980
206. Kellenberger E, Springael J-Y, Parmentier M, Hachet-Haas M, Galzi J-L, Rognan D (2007) J Med Chem 50:1294
207. Liao C, Karki RG, Marchand C, Pommier Y, Nicklaus MC (2007) Bioorg Med Chem Lett 17:5361
208. Lu Y, Nikolovska-Coleska Z, Fang X, Gao W, Shangary S, Qiu S, Qin D, Wang S (2006) J Med Chem 49:3759
209. Rella M, Rushworth CA, Guy JL, Turner AJ, Langer T, Jackson RM (2006) J Chem Inf Model 46:708
210. Saario SM, Poso A, Juvonen RO, Jaervinen T, Salo-Ahen OMH (2006) J Med Chem 49:4650
211. Tintori C, Manetti F, Veljkovic N, Perovic V, Vercammen J, Hayes S, Massa S, Witvrouw M, Debyser Z, Veljkovic V, Botta M (2007) J Chem Inf Model 47:1536

212. Barreca ML, De Luca L, Iraci N, Rao A, Ferro S, Maga G, Chimirri A (2007) J Chem Inf Model 47:557
213. Das K, Clark AD Jr, Lewi PJ, Heeres J, De Jonge MR, Koymans LM, Vinkers HM, Daeyaert F, Ludovici DW, Kukla MJ, De Corte B, Kavash RW, Ho CY, Ye H, Lichtenstein MA, Andries K, Pauwles R, Debethune M-P, Boyer PL, Clark P, Hughes SH, Janssen PA, Arnold E (2004) J Med Chem 47:2550
214. Murray-Rust J, Leiper J, McAlister M, Phelan J, Tilley S, Santa Maria J, Vallance P, McDonald N (2001) Nat Struct Biol 8:679
215. Congreve M, Chessari G, Tisi D, Woodhead AJ (2008) J Med Chem 51:3661
216. Congreve M, Murray CW, Carr R, Rees DC (2007) Annu Rep Med Chem 42:431
217. Hajduk PJ, Greer J (2007) Nat Rev Drug Disc 6:211
218. Hajduk PJ, Huth JR, Sun C (2006) Methods Princ Med Chem 34:181
219. Hajduk PJ (2006) Mol Interventions 6:266
220. Batista J, Bajorath J (2007) J Chem Inf Model 47:59
221. Crisman TJ, Sisay MT, Bajorath J (2008) J Chem Inf Model 48:1955
222. Chemical_Computing_Group http://www.chemcomp.com
223. Rummey C, Nordhoff S, Thiemann M, Metz G (2006) Bioorg Med Chem Lett Letters 16:1405
224. Krallinger M, Erhardt RA-A, Valencia A (2005) Drug Disc Today 10:439

Top Med Chem (2009) 5: 125–139
DOI: 10.1007/7355_2009_5

Published online: 30 June 2009

NMR Spectroscopy in Fragment Based Drug Design

Maurizio Pellecchia

Abstract In this chapter we will briefly reiterate critical aspects of solution nuclear magnetic resonance (NMR) spectroscopy approaches and their applications in drug discovery. These approaches are essentially based on a number of NMR techniques that have been developed to monitor and characterize intermolecular interactions. By way of examples, we will illustrate the unique advantages that these techniques offer when employed in conjunction with fragment-based ligand design, especially when tackling challenging drug targets.

Keywords NMR, drug discovery, fragment based drug discovery, SAR by NMR, SAR by ILOEs, hit to lead

Contents

1 Introduction

In the realm of drug discovery, the detection and characterization of target–ligand interactions is a critical process for lead identification, validation, and optimization.

M. Pellecchia(✉)
Burnham Institute for Medical Research, 10901 North Torrey Pines Road, La Jolla, CA 92037, USA
e-mail: mpellecchia@burnham.org

In the past decade a number of nuclear magnetic resonance (NMR) approaches have been proposed for such purposes [1–3]. These techniques generally exploit the fact that measurable NMR parameters of either a drug target (a protein or a nucleic acid) or a ligand differ depending on whether the molecules are isolated in solution or form a complex. Such techniques can be divided into two main categories, including a "direct" approach, where the target NMR observables are directly monitored in presence of a saturating concentration of a small molecule binder, and a second, indirect approach, in which the changes in nuclear spin relaxation of a small molecule are measured in presence of a substoichiometric amount of target. A third category of NMR experiments deals with the detection of enzyme kinetics and inhibition, in which the consumption or formation of a substrate or a cofactor is directly monitored [4, 5].

The major advantages of these strategies in the drug discovery process is that the intervening techniques are less prone to false positives and false negatives than any other type of assay and that there is no need to develop a specific assay for each target to monitor ligand binding. A major drawback of these approaches is their relatively low throughput and the relatively high amounts of target needed per assay, compared to spectrophotometric assays. These considerations make NMR spectroscopy the method of choice for hit validation, in which compounds that are identified in other assays, usually from high-throughput screening (HTS) campaigns, are verified for their binding to the target [3, 6]. However, because NMR techniques are exquisitely sensitive and enable the unambiguous detection of even weak binding events (with dissociation constants in the millimolar range), these methods also found wide applications in fragment-based drug discovery approaches (FBDD) [1, 2, 7, 8], in which high affinity binders for a given target are built stepwise from initial weakly interacting low molecular weight (MW) compounds (fragments). Because these approaches are independent from the nature of the target (protein or nucleic acids), it is obvious that the NMR techniques reiterated in this chapter will play an increasingly important role in chemical biology and in several aspects drug discovery from hit validation, to identification and optimization, particularly when tackling less characterized targets or those for which HTS techniques have failed to produce viable leads. The next paragraphs will reiterate the basic concepts of fragment based drug design, featuring the unique capabilities that NMR spectroscopy has to offer in these endeavors.

2 Fragment-Based Ligand Design: Puzzling Approaches to Drug Discovery

In a HTS campaign, typically hit compounds are ranked as a function of the number of initial positives (varying from 0.1 to 0.001% of the screened compounds), followed by those for which dose responses could be obtained (generally only 10% or less of the original hits). Of the few remaining compounds (if any), follow-up experiments and secondary assays are performed that serve to validate further the hits suitable for further medicinal chemistry efforts in the hit-to-lead optimization process. Considering that it has been estimated that the number of theoretically possible drug molecules is of the order of 10^{10}–10^{50} [9], it is obvious that while screening hundreds of thousands – or even millions – of compounds increases the chances of discovering valuable hits that can be optimized into *leads*, the process of validating hundreds or even thousands of hit compounds from an HTS campaign may be at odds with the initial intent of exploring a larger chemical space. The main issue with screening larger and larger libraries in HTS is the quite common risk that eventual valuable compounds may be "buried" in the inevitable large ensemble of false positives. In fact, from a recent survey summarizing a 900-page HTS industry account, representing data from 58 laboratories and 34 suppliers reports some interesting trends about the integration and rate of success of HTS campaigns [10], laboratories equipped for performing ultra HTS (uHTS; screening >100,000 compounds per week) were *not* proportionally more successful in generating *lead* compounds than laboratories with medium-throughput capabilities or even academic and noncommercial laboratories (testing only 3,000–7,000 compounds per week). Furthermore, when dealing with larger libraries it is not practically attainable to prefilter compounds for considerations related to ADME-T (absorption, distribution, metabolism, excretion and toxicity) liabilities for the use of the mature leads (and later drugs) in vivo, which is a common flaw in the HTS derived leads for drug development. One school of thought in recent years has been to filter the databases to be screened such that criteria largely comply with Chris Lipinski's "Rule of Five," an empirical list based on clinically successful drugs with good bioavailability [11, 12]. In essence, ligands with MW much larger than about 500 Da, flexible in nature (more than five rotatable binds), with a number of hydrogen bond donors and acceptors larger than ten, and with poor solubility, tend to become problematic at the developmental stages, especially for drugs intended for oral administration [13].

These problems tend to worsen when dealing with targets involving protein–protein interaction sites, for which to date HTS campaigns have generally failed to provide viable leads. Nevertheless, it is not completely clear if the screening campaigns are not producing valuable leads against protein–protein interactions because of the supposedly "*undrugable*" nature of the larger binding surfaces involved or simply because the compound libraries employed are highly populated by compounds that were not originally derived to mimic or complement a protein surface. With these considerations in mind, it should be intuitive that a more rational chemical design approach, in which a molecule is iteratively "built" stepwise within the binding cavity of a given target, would likely represent a more successful strategy, especially when dealing with drug targets presenting larger surfaces and/or that have resulted very challenging for HTS.

These simple considerations form the basis for FBDD [7, 8] that, in essence, have the intrinsic advantage of exploring in principle a much larger accessible chemical space than conventional HTS campaigns, by rationally designing high affinity binders from a small but diverse set of building blocks (fragments).

In essence, the strategy consists of building up a lead compound (with a typical MW of less than 500 Da) from screening small libraries (typically 1,000–15,000 compounds) composed of smaller molecules (fragments) with MW <300 Da and good aqueous solubility. The most common FBDD approaches include tethering [7, 8, 14–27], X-ray diffraction [22, 28–33], or NMR spectroscopy (reiterated here) [1, 7, 8, 26, 34–40] as methods for fragment screening and to guide iterative optimizations. More recently, applications involving surface plasmon resonance have also been proposed [41].

Fundamentally, FBDD is a general approach aimed at deriving high affinity ligands for macromolecular targets starting from low MW binders, which are usually identified by the use of the aforementioned biophysical techniques [1, 2, 7, 8].

Following the identification of initial weak binders, two general approaches are adopted to increase the affinity of the ligands for the given target, the most straightforward being testing analogs with increased MW and hence chemical complexity (Fig. 1). This approach is usually straightforward as several compound analogs are usually commercially available. Subsequently, the evolution of the resulting hit can follow structure-based refinements of the initial chemical structure based on either a computational model or, as is most often the case, by an experimentally derived (by using X-ray or NMR spectroscopy) structure of the complex. Obviously, an intermediate approach consists of designing a new compound that has chemical substructures from two fragments, that occupy partially overlapping subsites, "merged" into a new molecule. These compounds can subsequently serve as the starting point for more traditional medicinal chemistry and structure–activity relationships based studies aimed at the optimization of binding affinity and selectivity of the compounds. One important consideration is that these approaches enable a better control of the nature of the interactions, hence avoiding false positives, and also provides the opportunity for keeping drug-likeness criteria of the compounds in check. Because the compounds constituting the fragment libraries are "building blocks" of the final hits, rather than the traditional Lipinski rule-of-five [11, 12], a more stringent rule-of-three has been proposed [15]. Hence, the fragment libraries are made out of scaffolds or fragments that per se do not in principle contain chemical substructures or possess physical characteristics that cause undesired properties and may lead to adverse effects in vivo. Careful selection of the fragment libraries is therefore pivotal for the success of these approaches. Of note is that the small size of these libraries renders them more practically and economically amenable to such careful selection, a task that is not as easily attainable when dealing with the very large libraries used for HTS or uHTS, that will inevitably be populated by undesirable compounds, from a subsequent lead optimization and even drug development points of view. Finally, one should also note that most approved drugs tend to contain a small variety of common chemical substructures, as if there are certain scaffolds that have a better chance to interfere with biological targets and to be safe and effective for use in humans. Hence, fragment libraries populated in these chemical scaffolds could lead to more drug-like mature lead compounds [42].

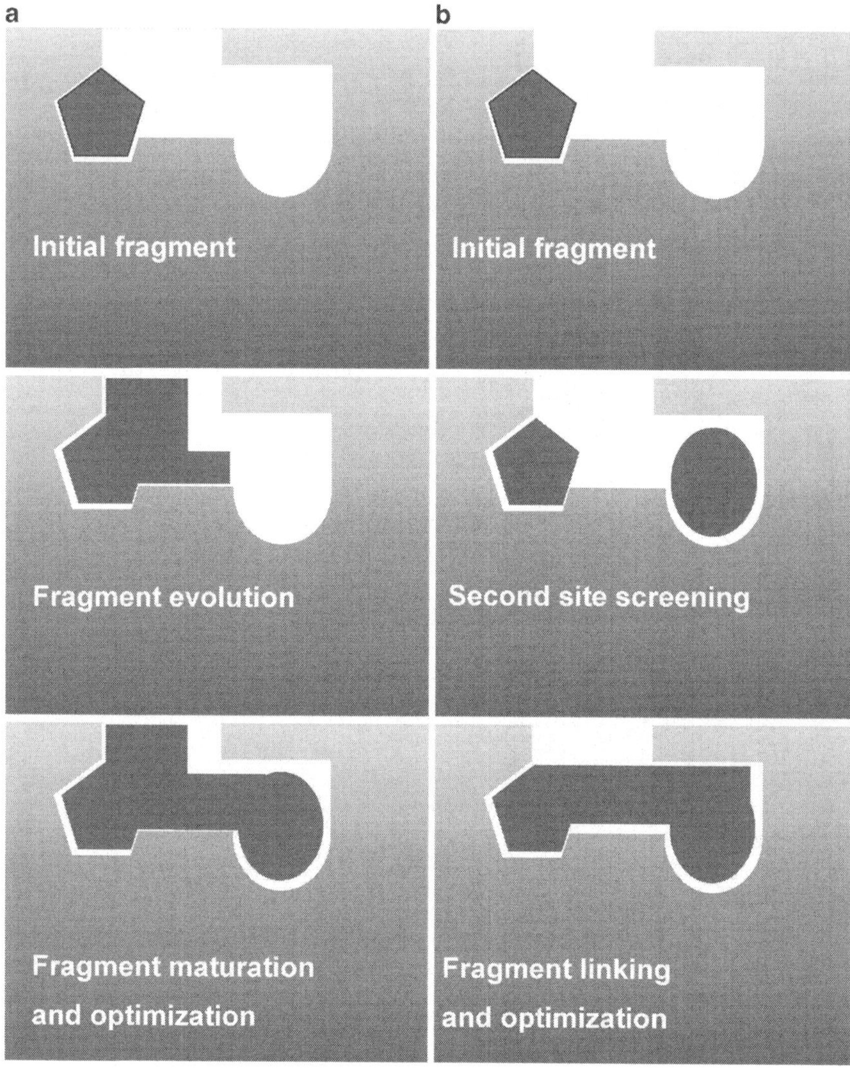

Fig. 1 Schematic representations of the two general FBDD approaches. (**a**) Evolution of the fragment in the binding site of a generic target. (**b**) Fragment-linking approach

A more sophisticated FBDD approach consists of designing bidentate compounds chemically linking two weakly interacting scaffolds that occupy adjacent subpockets on the target's surface (Fig. 1b). In this case, the free energy of binding of the resulting bidentate compound with respect to those of the individual fragments can be expressed as

$$\Delta G^{\mathrm{AB}} = \Delta H^{\mathrm{A}} + \Delta H^{\mathrm{B}} - -\mathrm{T}\Delta S^{\mathrm{AB}} = -RT\ln\left(K_{\mathrm{D}}^{\mathrm{A}} K_{\mathrm{D}}^{\mathrm{B}} E\right),$$

where R represents the Boltzman constant, T is the temperature of the system, ΔH^{A} and ΔH^{B} are the enthalpy of binding of fragments A and B respectively, ΔS^{AB} represents the entropy loss upon binding of the bidentate compound, $K_{\mathrm{D}}^{\mathrm{A}}$ and $K_{\mathrm{D}}^{\mathrm{B}}$ are the dissociation constants of the individual initial binders, and E is the linking coefficient [26]. Hence, linking even weak binders occupying adjacent pockets on the surface of the macromolecular target can result in high affinity bidentate compounds. In fact, in an ideal case, the $\Delta \mathrm{S}^{\mathrm{AB}}$ value would be approximately equal to the loss of entropy of binding of the isolated compounds (A or B). Therefore, the binding affinity of the bidentate compound would be, in principle, higher than that of the individual compounds not only because of the larger number of interactions (enthalpy factor), but also due to a reduced loss in entropy upon binding (factor E). The approach has been demonstrated to yield bidentate compounds with affinities dramatically improved with respect to those of the individual fragments [26].

The choice of the fragment evolution approach vs the fragment-linking approach depends on the nature of the target. For large surfaces, such as those involved in protein–protein interactions, for example, it is expected that the latter approach may be more appropriate. However, for smaller binding pockets, such as those for enzymes, for example, the fragment evolution could be more successful as there may not be sufficient space to accommodate two fragments simultaneously and/or to accommodate a linker region.

When choosing the fragments to be optimized in an FBDD campaign, a key parameter to rank order hits is the ligand efficiency (or binding efficiency index) [43–45], which is defined as the free energy of binding per nonhydrogen atom. The ligand efficiency allows normalizing MW and potency of a given molecule, hence it provides a more rational and intuitive rank ordering of hit compounds [44, 45]. In successful fragment optimizations, the potency increases linearly with MW at an average rate of about 0.3 kcal mol^{-1} per atom. Hence, it is obvious that selecting initial hit fragments with the most optimal ligand efficiency is a prerequisite for the development of a potent lead molecule with acceptable MW [46].

The following paragraphs will report on basic NMR techniques that are highly relevant for hit validation and for FBDD applications.

3 Chemical Shift Perturbation and Related Methods

A simple NMR technique, and arguably the most widely used and effective for hit validation, is the chemical shift perturbation method. In this approach, a reference spectrum of isotopically labeled target is recorded in absence and presence of a given test ligand (or a mixture of test ligands). Commonly, differences in chemical shift between free and bound protein target are observed in 2D [^{15}N, ^{1}H] and/or 2D [^{13}C, ^{1}H] correlation spectra of a protein (or nucleic acid) upon titration of a ligand

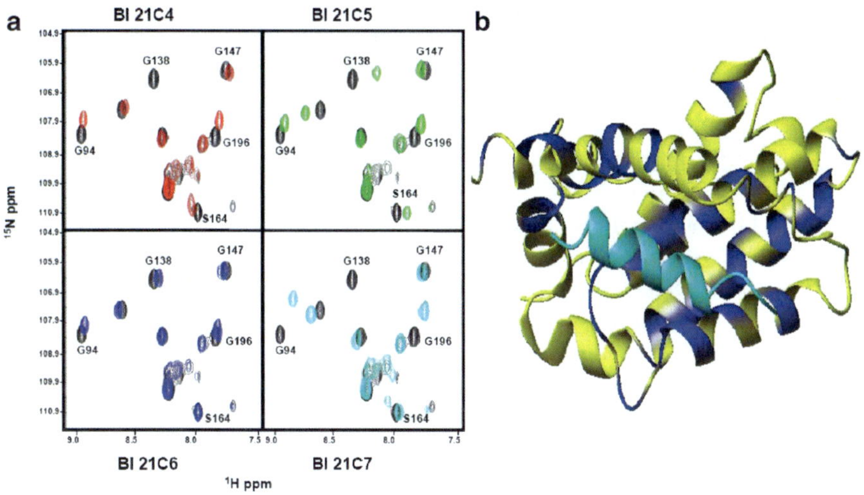

Fig. 2 Chemical shift perturbation and chemical shift mapping. (**a**) Portions of the [^{15}N, ^{1}H]-HSQC spectra of Bcl–x_L recorded in absence (*black*) and in presence of each of the four molecules (*in colors*). Resonance assignments for amino acid residues that exhibit large shifts are reported. (**b**) Structure of Bcl–x_L in complex with the BH3 peptide from Bak (PDB code 1BXL) showing the chemical shift changes in Bcl–x_L upon ligand binding (*blue*, large shits; *yellow*; no shifts; the Bak peptide is reported in *cyan*). Adapted from [48]

or a mixture of ligands, providing proof of binding and a value for the dissociation constant of the complex [26, 47]. Chemical shift perturbation studies can also provide crude structural information on the site of binding, when the resonance assignments are known [26, 47] (Fig. 2). This approach is known as chemical shift mapping. Nowadays collections of [^{15}N, ^{1}H] or [^{13}C, ^{1}H] correlations spectra (either TROSY-type [49] or conventional HMQC and HSQC) with uniformly or amino acid type selectively labeled protein samples [50] are the methods of choice for both chemical shift perturbation or chemical shift mapping studies [2].

The use of chemical shift perturbation to monitor ligand binding has several major advantages. First, binders to a given protein can be found without the need of developing a specific assay or even regardless of the knowledge of its function. This is generally true for most of NMR-based techniques, although the chemical shift mapping is superior to other NMR-based approaches as it also discriminates between specific and nonspecific binding. In fact, as mentioned, given that the resonance assignments and the three-dimensional structure of the target are known, the location of the site of binding can be obtained. The latter enables a crude yet rapid and efficient assessment of the site of binding for the ligand on the target's surface [50]. This information can be combined with molecular modeling studies to obtain rapidly a picture of the binding mode of the hit compound that can be used to formulate hypotheses for further optimizations [51]. In addition, if the structure of the target was previously determined by NMR, in some instances it is possible to derive rapidly intermolecular NOE-type constraints (that translate into internuclear distance measurements) to position more precisely the ligand on the protein binding site [26]. The drawback of the chemical shift mapping is that the amount of protein needed for a single NMR experiment is still relatively high for the technique to be used efficiently to test large libraries of compounds, and the fact that the protein needs to be labeled. However, the technique is also very sensitive to weak binding events, a fact that is common to most of NMR-based drug discovery strategies. Hence, when such a weak binder (a fragment) is found to bind to a given protein, follow-up hit optimizations strategies can be devised to increase iteratively the affinity of the compound for the given target [1].

In this respect, the most popular strategy is the SAR by NMR approach [26, 47], in which a chemical shift mapping-based screen for a second binder is performed in the presence of an initial weakly interacting first ligand. Compounds that induce chemical shift changes that correspond to a region on the protein surface that is adjacent to the site of binding of the first ligand are considered. The structural characterization of the ternary complex by NMR allows the design of potential chemical linkers between the compounds to afford a more potent ligand. In practical applications, initial bidentate compounds represent the starting point for traditional SAR-based (structure–activity relationships) optimizations to obtain potent and selective compounds (Fig. 3). This approach proved to be very useful against a variety of targets, including those for which other techniques did not produce viable lead compounds (see for example [52]).

4 Transferred NMR Measurements to Detect Ligand Binding and Related Methods

Without going deeper into technical details, nuclear spin relaxation is dominated by the rotational correlation time of the molecule. A ligand can assume quite different apparent rotational correlation times in solution whether it is free or bound to a large macromolecular target; hence its NMR observables will exhibit different relaxation properties. These relative differences can be exploited in detecting ligand binding and in characterizing structurally the ligand binding mode. In nontechnical terms, NMR spectroscopy measurements are dominated by kinetics of small molecule dissociation from the target relative to nuclear relaxation times for NMR observables (most often hydrogen nuclei). In the so-called "fast exchange regime" the relaxation properties of a ligand at the equilibrium with its complex state with a macromolecular target are the weighted average of the values corresponding to the free and fully bound states. This occurs when the off rate of the complex is fast compared with the differences in relaxation times of the complexed vs free ligand [2]. Moreover, in the fast exchange regime (as occurs for ligands with low-micromolar to millimolar binding affinities), binding can then be detected by measuring the relaxation properties of the test ligand in the presence of a substoichiometric amount of target [2].

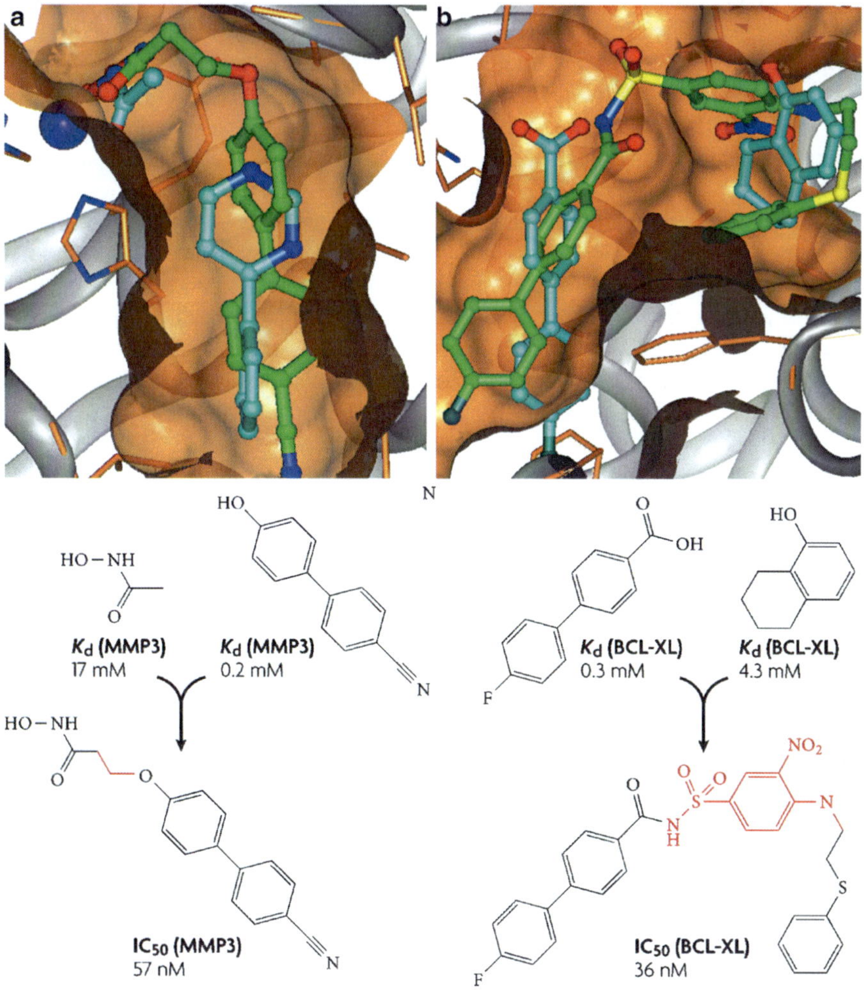

Fig. 3 The SAR by NMR approach. Example of a small bidentate molecule designed using this approach. The example shown is for the design of a potent inhibitors of the matrix metalloproteinase MMP3. (**a**) Docked structures of the identified fragment leads are shown with *cyan* carbons, whereas the linked compound is shown with *green* carbon atoms. All structures were experimentally determined by NMR. (**b**) Chemical structures (and in vitro potencies) of the fragment leads and subsequent high-affinity linked compounds. Adapted from [7]

A practical application consists of measuring transverse relaxation rates of a ligand in the presence of a substoichiometric amount of target. Hence, ligand screening can be achieved by measuring a 1D ^{1}H NMR spectrum of test ligand(s) with a relaxation filter (a spin-lock) in the presence and absence of a substoichiometric amount of target. The relaxation filter has two functions: first, it eliminates any residual signal from the target; second, it serves to enhance the differences in signal intensities in binders vs nonbinders. In typical applications, ligands can be tested as mixtures at concentrations as low as 50 μM in the presence of as low as 1 μM of target, respectively [53].

Other efficient applications involve a direct transfer of magnetization between the target and the bound ligand as a way to detect binding. The saturation transfer difference (STD) experiment [54, 55] is based on a selective irradiation of protein resonances, which is obtained by irradiating the regions of the ^{1}H NMR spectrum that are usually not occupied by resonances from small organic molecules, for example the aliphatic region of the spectrum, between −1 and 2 ppm. The irradiation is transferred to the bound ligand via the Nuclear Overhauser Effect, and is manifested in a decrease of signal intensity of the bound ligand compared to a reference spectrum in which the irradiation is outside the spectral region [54, 55]. Hence, in a mixture of test ligands, subtracting the reference spectrum from the protein irradiated spectrum will result in the NMR spectrum of binding compounds. WaterLOGSY (Water ligand observation by gradient spectroscopy) is a related experiment in which the selective saturation of the protein is achieved indirectly by irradiation of water protons [56]. These experiments exploit simple 1D NMR experiments and can be useful in screening as well as in hit validation and in some instances to provide a crude measure of the dissociation constants or displacement constants [57, 58].

In the STD experiment, the epitope of the interactions can be obtained from the relative saturation of the resonance lines in within a given binder. This is particularly useful and applicable for large ligands such as peptides or carbohydrates [59, 60]. More detailed structural constraints on the bioactive conformation of larger ligands can also be obtained via transferred 2D [^{1}H, ^{1}H]-NOESY spectra, measured in the presence of a small amount of protein target. These experiments provide distance-dependant intramolecular NOEs signals which can be used to determine the bioactive conformation of the ligand [61].

In addition to intramolecular NOEs, the observation of protein mediated ligand–ligand NOEs (ILOEs) can be extremely informative in the design of potential bidentate compounds and for lead optimization [2, 42, 62, 63]. Hence, ILOEs between adjacent compounds are only mediated by the protein that is therefore needed only in small amounts and unlabeled (Fig. 4). Much as for all experiments that are based on the observation of the ligands, the size of the protein is not a limiting factor. On the contrary, it is common experience that the larger the protein, the more efficient cross-relaxation (that generates the NOEs) between the adjacent ligands becomes [3].

In an application called SAR by ILOE [62, 63, 65], in analogy to the SAR by NMR method, the design of potential bidentate compounds can be achieved via detection of ILOEs in compound mixtures from small but diverse libraries of low MW compounds (fragments). In a related approach, the chemical nature of the identified pair of fragments can be used as a starting point for a pharmacophore based design of potential bidentate compounds (Fig. 4) [64]. This has the advantage that compounds could be rapidly and directly selected for testing from large libraries of commercially available molecules before initiating synthetic chemistry (Fig. 4).

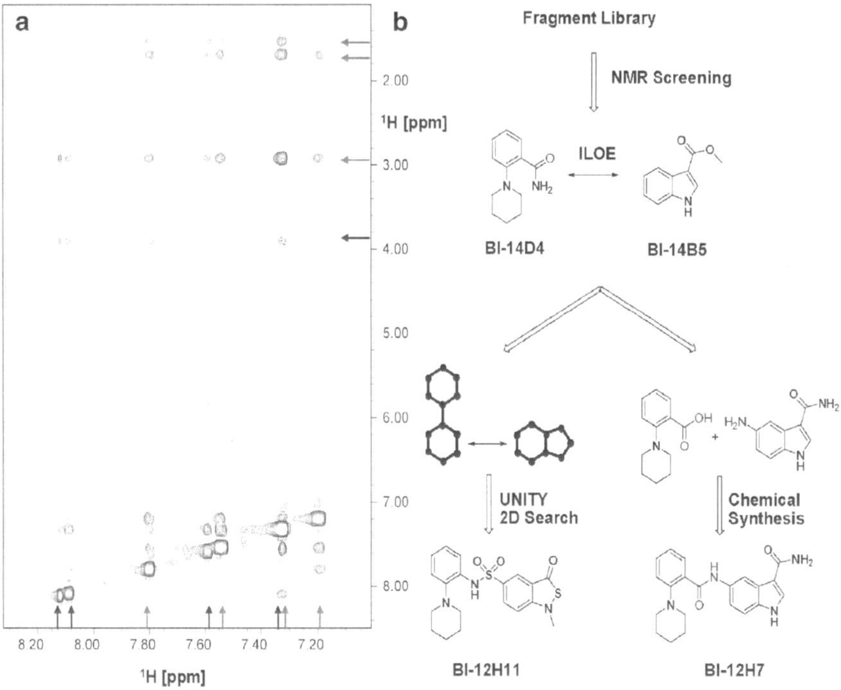

Fig. 4 Protein-mediated ligand–ligand magnetization transfer. (**a**) Interligand NOEs in the aliphatic region of two weak-binding fragments reported from a screening mixture. Intra- and interligand NOE cross peaks between the weakly interacting ligands (500 μM concentration each) are detected in a 2D [^{1}H, ^{1}H] NOESY experiment, in presence of a substoichiometric amount of p38α (5 μM). The NOE cross-peaks of the reported ligands are labeled with *green (left compound)* and *blue (right compound) arrows*, respectively. (**b**) Scheme for the discovery of p38α inhibitors from a pair of binding fragments by pharmacophore search (Pharmacophore by ILOEs) and by chemical synthesis (SAR by ILOEs) of a high affinity bidentate compound. Adapted from [64]

Finally, an approach that is very efficient in artificially enhancing the transverse relaxation rates of a ligand upon binding is to introduce a paramagnetic spin on the target. For example, if the target contains a metal-ion in proximity or within the binding site, the paramagnetic center could simply be the metal ion itself (e.g., Mn^{2+}) [50]. For other targets, a paramagnetic spin can be artificially introduced. A common spin-label is 2,3,4,6-tetramethyl-piperidine-1-oxyl (known as TEMPO) [66]. A very useful application consists of utilizing the paramagnetic approach for a second site screen in hit optimization projects [67]. This could be obtained by designing a first ligand labeled with a spin-label, hence exploiting the relaxation enhancement effect that the compound may induce to compounds that bind on the surface of the target in proximity to the first molecule. In analogy to the SAR by NMR and SAR by ILOE approaches, this strategy would lead to a pair of compounds that occupy adjacent sites on the protein surface (Fig. 5). While the drawback of the method is

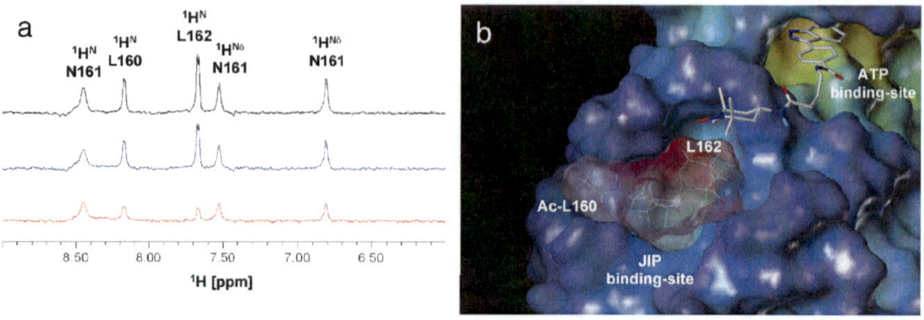

Fig. 5 Paramagnetic spin-labeled compounds for second site screening. Example of screening for peptide binding in the docking site of the protein kinase JNK. An indazole ATP mimic has been derivatized with a TEMPO molecule that brings a paramagnetic unpaired electron just at the edge of the docking site of JNK. Hence, ligand binding in this region can be detected by monitoring the signal intensity of a test ligand in presence of protein and the reference indazole-TEMPO molecule. (**a**) 1D NMR spectra of the test peptide Ac–LNL–OH (1 mM) are measured in presence of 200 μM of the TEMPO compound (*black*), in presence of 10 μM of protein target (*blue*) (JNK), and in presence of both 200 μM of TEMPO and 10 μM of protein JNK (*red*). (**b**) Docking of the indazole-TEMPO molecule and peptide Ac–LNL–OH in JNK. The surface of the protein is displayed to highlight the cavities, the compound is displayed in capped sticks without protons to better visualize the structure and the peptide is displayed in sticks with a translucent surface in a gradient from *red* to *gray* coding for the effect of the paramagnetic probe on the resonance lines of the peptide, as shown in **a**: *red*, more affected; *gray*, less affected. Peptide pose corresponds to that obtained directly from the X-ray structure of the complex (PDB-ID 1UKI). Adapted from [68]

that a spin-labeled compound has to be obtained [67, 69], the advantage of this method is that the protein is simply mediating the interactions. Hence, only small amounts of unlabeled target are needed and the size of the target is not a limiting factor.

5 Conclusions and Outlook

NMR has found an increasingly important role in the study and characterization of the interactions between small organic molecules and macromolecular targets, which is particularly useful in drug discovery [35]. These approaches provide a clear cut validation for hit compounds arising from HTS campaigns. The use of [^{15}N,^{1}H] correlation spectra or relatiation measurements are becoming routine experiments to demonstrate a direct interaction between a hit compound and its macromolecular target [3]. In addition, NMR strategies to support fragment-based discovery and optimization have emerged in the past 5–10 years that are opening the way to the design of compounds against challenging drug targets [7, 8]. Hence, we anticipate a significant role for NMR-based techniques in the near future, both at the early stages of hit identification and validation, and to develop small organic molecules capable of modulating unconventional drug targets such as those involved in protein–protein interactions or even nucleic acids and membrane proteins.

Acknowledgments Several review articles cover in much greater details numerous NMR applications that have been reported in the past years on the use of NMR spectroscopy in the drug discovery process and to monitor target–ligand interactions. The author apologizes if he was unable to mention properly all this work within the limited space of this article.

References

1. Pellecchia M, Sem DS, Wuthrich K (2002) Nat Rev Drug Discov 1:211
2. Pellecchia M (2005) Chem Biol 12:961
3. Pellecchia M, Becattini B, Crowell KJ, Fattorusso R, Forino M, Fragai M, Jung D, Mustelin T, Tautz L (2004) Expert Opin Ther Targets 8:597
4. Papeo G, Giordano P, Brasca MG, Buzzo F, Caronni D, Ciprandi F, Mongelli N, Veronesi M, Vulpetti A, Dalvit C (2007) J Am Chem Soc 129:5665
5. Dalvit C, Ardini E, Flocco M, Fogliatto GP, Mongelli N, Veronesi M (2003) J Am Chem Soc 125:14620
6. Hajduk PJ, Burns DJ (2002) Comb Chem High Throughput Screen 5:613
7. Hajduk PJ, Greer J (2007) Nat Rev Drug Discov 6:211
8. Jahnke W, Erlanson DA (2006) Fragment-based approaches in drug discovery. Wiley, New York
9. Valler MJ, Green D (2000) Drug Discov Today 5:286
10. Vogtherr M, Saxena K, Hoelder S, Grimme S, Betz M, Schieborr U, Pescatore B, Robin M, Delarbre L, Langer T, Wendt KU, Schwalbe H (2006) Angew Chem Int Ed Engl 45:993
11. Lipinski CA, Lombardo F, Dominy BW, Feeney PJ (2001) Adv Drug Deliv Rev 46:3
12. Lipinski CA (2003) Drug Discov Today 8:12
13. Lipinski CA (2000) J Pharmacol Toxicol Methods 44:235
14. Carr RA, Congreve M, Murray CW, Rees DC (2005) Drug Discov Today 10:987
15. Congreve M, Carr R, Murray C, Jhoti H (2003) Drug Discov Today 8:876
16. Erlanson DA, Hansen SK (2004) Curr Opin Chem Biol 8:399
17. Erlanson DA, McDowell RS, O'Brien T (2004) J Med Chem 47:3463
18. Erlanson DA, Wells JA, Braisted AC (2004) Annu Rev Biophys Biomol Struct 33:199
19. Gill A, Cleasby A, Jhoti H (2005) Chembiochem 6:506
20. Hartshorn MJ, Murray CW, Cleasby A, Frederickson M, Tickle IJ, Jhoti H (2005) J Med Chem 48:403
21. Moore WR Jr (2005) Curr Opin Drug Discov Devel 8:355
22. Rees DC, Congreve M, Murray CW, Carr R (2004) Nat Rev Drug Discov 3:660
23. Schade M, Oschkinat H (2005) Curr Opin Drug Discov Devel 8:365
24. Villar HO, Yan J, Hansen MR (2004) Curr Opin Chem Biol 8:387
25. Zartler ER, Shapiro MJ (2005) Curr Opin Chem Biol 9:366
26. Shuker SB, Hajduk PJ, Meadows RP, Fesik SW (1996) Science 274:1531
27. Langer T, Vogtherr M, Elshorst B, Betz M, Schieborr U, Saxena K, Schwalbe H (2004) Chembiochem 5:1508
28. Allen KN, Lavie A, Petsko GA, Ringe D (1995) Biochemistry 34:3742
29. Rutenber E, Fauman EB, Keenan RJ, Fong S, Furth PS, Ortiz de Montellano PR, Meng E, Kuntz ID, DeCamp DL, Salto R et al. (1993) J Biol Chem 268:15343
30. Rutenber EE, McPhee F, Kaplan AP, Gallion SL, Hogan JC Jr, Craik CS, Stroud RM (1996) Bioorg Med Chem 4:1545
31. Shoichet BK, Stroud RM, Santi DV, Kuntz ID, Perry KM (1993) Science 259:1445
32. Verlinde CLMJ, Pijning T, Kalk KH, van Calenbergh S, Aershot AV, Herdewijn P, Callens M, Michels P, Opperdoes FR, Wierenga RK, Hol WGJ (1993) In: Testa B, Kyburz E, Fuhrer W, Giger R (eds) Perspectives in medicinal chemistry. Helvetica Chimica Acta, Basel, p 135

33. Bertini I, Del Bianco C, Gelis I, Katsaros N, Luchinat C, Parigi G, Peana M, Provenzani A, Zoroddu MA (2004) Proc Natl Acad Sci USA 101:6841
34. Hajduk PJ, Gerfin T, Boehlen JM, Haberli M, Marek D, Fesik SW (1999) J Med Chem 42:2315
35. Jahnke W (2007) J Biomol NMR 39:87
36. Kessler H, Klages J (2006) In: Triggle DJ, Taylor JB (eds) Comprehensive medicinal chemistry II, vol 3. Elsevier, Oxford, p 901
37. Klages J, Coles M, Kessler H (2006) In: Bartlett PA, Etzeroth M (eds) NMR-based screening in exploiting chemical diversity for drug discovery, vol 12. RSC, Cambridge, UK, p 263
38. Klages J, Coles M, Kessler H (2006) NMR-based screening. RSC, Cambridge, UK
39. Klages J, Coles M, Kessler H (2007) Analyst 132:693
40. Coles M, Heller M, Kessler H (2003) Drug Discov Today 8:803
41. Neumann T, Junker HD, Schmidt K, Sekul R (2007) Curr Top Med Chem 7:1630
42. Fejzo J, Lepre CA, Peng JW, Bemis GW, Ajay, Murcko MA, Moore JM (1999) Chem Biol 6:755
43. Kuntz ID, Chen K, Sharp KA, Kollman PA (1999) Proc Natl Acad Sci USA 96:9997
44. Hopkins AL, Groom CR, Alex A (2004) Drug Discov Today 9:430
45. Cele AZ, Metz JT (2005) Drug Discov Today 10:464
46. Hajduk PJ (2006) J Med Chem 49:6972
47. Hajduk PJ, Sheppard G, Nettesheim DG, Olejniczak ET, Shuker SB, Meadows RP, Steinman DH, Carrera GM, Marcotte PA, Severin J, Walter K, Smith H, Gubbins E, Simmer R, Holzman TF, Morgan DW, Davidsen SK, Summers JB, Fesik SW (1997) J Am Chem Soc 119:5818
48. Rega MF, Leone M, Jung D, Cotton NJ, Stebbins JL, Pellecchia M (2007) Bioorg Chem 35:344
49. Pervushin K, Riek R, Wider G, Wuthrich K (1997) Proc Natl Acad Sci USA 94:12366
50. Pellecchia M, Meininger D, Dong Q, Chang E, Jack R, Sem DS (2002) J Biomol NMR 22:165
51. Wyss DF, McCoy MA, Senior MM (2002) Curr Opin Drug Discov Devel 5:630
52. Oltersdorf T, Elmore SW, Shoemaker AR, Armstrong RC, Augeri DJ, Belli BA, Bruncko M, Deckwerth TL, Dinges J, Hajduk PJ, Joseph MK, Kitada S, Korsmeyer SJ, Kunzer AR, Letai A, Li C, Mitten MJ, Nettesheim DG, Ng S, Nimmer PM, O'Connor JM, Oleksijew A, Petros AM, Reed JC, Shen W, Tahir SK, Thompson CB, Tomaselli KJ, Wang B, Wendt MD, Zhang H, Fesik SW, Rosenberg SH (2005) Nature 435:677
53. Hajduk PJ, Olejniczak ET, Fesik SW (1997) J Am Chem Soc 119:12257
54. Klein J, Meinecke R, Mayer M, Meyer B (1999) J Am Chem Soc 121:5336
55. Mayer M, Meyer B (1999) Angew Chem Int Ed 38:1784
56. Dalvit C, Pevarello P, Tato M, Veronesi M, Vulpetti A, Sundstrom M (2000) J Biomol NMR 18:65
57. Dalvit C, Flocco M, Stockman BJ, Veronesi M (2002) Comb Chem High Throughput Screen 5:645
58. Dalvit C, Hadden DT, Sarver RW, Ho AM, Stockman BJ (2003) Comb Chem High Throughput Screen 6:445
59. Mayer M, Meyer B (2001) J Am Chem Soc 123:6108
60. Meinecke R, Meyer B (2001) J Med Chem 44:3059
61. Leone M, Freeze HH, Chan CS, Pellecchia M (2006) Curr Drug Discov Technol 3:91
62. Becattini B, Pellecchia M (2006) Chemistry 12:2658
63. Becattini B, Sareth S, Zhai D, Crowell KJ, Leone M, Reed JC, Pellecchia M (2004) Chem Biol 11:1107
64. Chen J, Zhang Z, Stebbins JL, Zhang X, Hoffman R, Moore A, Pellecchia M (2007) ACS Chem Biol 2:329
65. Becattini B, Culmsee C, Leone M, Zhai D, Zhang X, Crowell KJ, Rega MF, Landshamer S, Reed JC, Plesnila N, Pellecchia M (2006) Proc Natl Acad Sci USA 103:12602

66. Jahnke W, Rudisser S, Zurini M (2001) J Am Chem Soc 123:3149
67. Jahnke W, Lawrence B, Perez L, Paris G, Strauss A, Fendrich G, Nalin CM (2000) J Am Chem Soc 122:7394
68. Vazquez J, De SK, Chen LH, Riel-Mehan M, Emdadi A, Cellitti J, Stebbins JL, Rega MF, Pellecchia M (2008) J Med Chem 51:3460
69. Vazquez J, Tautz L, Ryan JJ, Vuori K, Mustelin T, Pellecchia M (2007) J Med Chem 50:2137

Top Med Chem (2009) 1: 141–174
DOI: 10.1007/7355_2009_1

Published online: 30 June 2009

Hit Triage: Medicinal Chemistry Strategies to Improve the Odds of Success in Discovery

Kevin D. Freeman-Cook and Daniel W. Kung

Abstract This chapter describes the process of collecting and analyzing information in order to make decisions which help a project move through a key early stage in drug discovery. It explains, in general, the sources of hits for a drug discovery program and explores the link between potency and efficiency in evaluating lead matter. The chapter then illustrates the collection of both pharmacokinetic data as well as safety data, and describes the relevance of that data and the process of using it to make decisions. Of course, there is no one "right" or "wrong" way to do the work contained in this chapter, but the practice of efficiently building a network of information on compounds, and then using that to make informed decisions can help to shift the odds of success in drug discovery.

Keywords Efficiency, Hit, Safety, Series, Triage

Contents

K.D. Freeman-Cook(✉)
9 Michael Drive, Clinton, CT 06413, USA
e-mail: Kevin.Freeman-Cook@pfizer.com

D.W. Kung
289 Laurelwood Drive, Salem, CT 06420, USA
e-mail: Daniel.W.Kung@pfizer.com

1 Introduction: Philosophy of "Hit Triage"

The identification of chemical matter with suitable structure and properties to lead to the successful identification of drug candidates and drugs is a critical part of drug discovery. In this chapter, "hit triage" is addressed. We consider hit triage to be a decision-making process that spans a range of activities: (1) evaluation and assessment of the known and expected properties of validated hit compounds, (2) description of the key issues facing progression of any hit(s), which helps to define the intermediate goals or questions to be addressed in hit follow-up, and (3) prioritization of hit compounds or series for further chemistry optimization [1]. The goal of hit triage is to evaluate and identify chemical matter that will enable the project to reach stated decision points (such as terminate chemistry, demonstrate preclinical proof-of-concept, identify a drug).

A major challenge in evaluation of hit matter is typically the relative lack of information that is available on the properties of the hit compounds. As compound series are progressed from hits to leads to drug candidates to drugs, the quantity of data that is available on a given compound increases dramatically, thereby increasing the confidence in the decision to progress or terminate study of a particular compound. However, at the beginning of a chemistry program (hit evaluation or hit triage), many more judgments must be made based on probabilities or extrapolation rather than knowledge.

A hit triage process is advocated here that balances (1) using data and knowledge to make decisions when possible, with (2) using general principles and hypotheses to assess probabilities in the absence of data. When linked together, the use of available data in conjunction with general principles provides a framework for defining the key questions to be addressed in the optimization of a hit compound series, as well as the medicinal chemistry strategy and tactics for that optimization.

1.1 Sources of Hits

The potential sources of the compounds that are evaluated in a hit triage process are described in other chapters in this volume. These sources may include a high throughput screen (HTS) of a diverse compound collection, a targeted screen of a gene-family biased set of compounds (e.g., kinase or protease), a screen of low molecular weight (MW) fragments, a targeted screen based on some extant knowledge (e.g., biased by virtual screening to increase hit rate), or literature reports of active compounds. Compounds from each of these sources of hits are likely to have available different sets of data and knowledge that will be important inputs to hit triage decision-making [2].

1.2 The Balance Between Data, Knowledge, and Probability

To illustrate the balance of data and knowledge vs probability, consider hits from a fragment screen. These hits will be low molecular weight (<300) and are often weak in terms of absolute potency (perhaps hundreds of micromolar); significant changes to the compound structure including an increase in molecular weight are likely to be necessary to achieve target potency. In the situation where substantial structural changes will be made, the hit structure provides very limited data or knowledge that will link directly to the properties of the optimized compounds [3]. From a decision-making standpoint, however, the power of the fragment approach is that it enables the medicinal chemist to make assessments of the probability that any given structural change to be introduced will impart desirable or undesirable compound properties. For the analysis of a series of compounds or multiple series of compounds, a probability-based analysis is expected to have the greatest utility. An obvious weakness in proceeding on the basis of probability, however, is the gap in correlating probability to actual experimental outcome for any single compound.

At the other end of the data and knowledge spectrum, hits from the literature can provide a wealth of experimental data and knowledge upon which to base decisions. Because these compounds have already undergone significant optimization, their data are likely very relevant to the properties that might arise from further optimization. In addition to reported biological data and structure activity relationships, these hits can be synthesized and characterized in any available assays prior to deciding how to proceed with medicinal chemistry plans. The potency, pharmacokinetic, safety, and intellectual property features can often be well understood, leading to comprehensive and well defined goals that need to be achieved, as well as identified assays with which to assess those properties. The number of unknown properties may be quite small. The existence and identification of assays to assess known liabilities can enable a much more empirically based setting of medicinal chemistry strategy, rather than a probability-based one (Fig. 1).

2 Properties to Consider

The identification of a drug requires optimization of the balance among multiple properties, among them potency/efficacy, pharmacokinetics (PK), safety, and intellectual property. As a consequence, these are also the factors that are important to

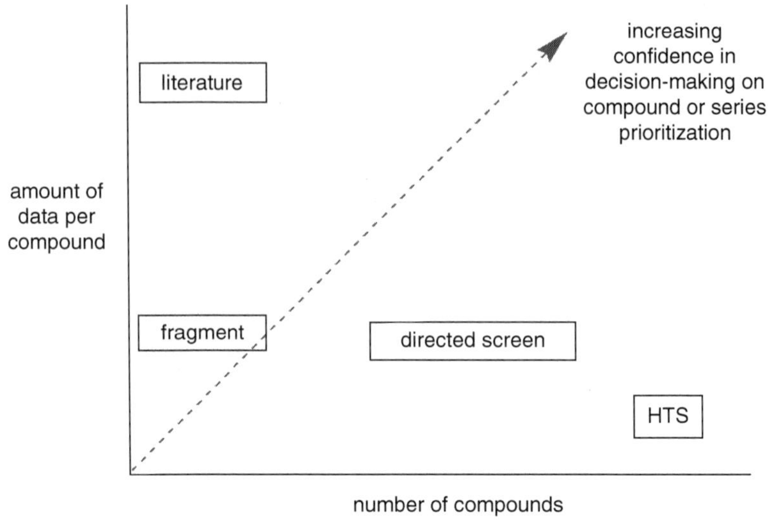

Fig. 1 Different sources of hits typically provide different distributions of number of hit compounds and amount known (and relevant) about those compounds, each of which contributes to the decision making process of hit selection and hit triage

consider for hit triage (assuming that the ultimate goal is to identify a drug – the considerations may be different for a preclinical tool compound, for example). The sections that follow on potency, PK, and safety are intended to provide a framework for considering the experimental data and assays that may be useful for evaluation of hits, as well as general principles on how probability assessments can be made when data are not available. Ultimately, the scientists working on a given project must decide how to balance multiple properties and probabilities. The decisions made about the properties, goals, probabilities, and prioritization for any hit compound series then lead directly into the execution of chemistry hit follow-up, which is the subject of this chapter.

The value of optimizing multiple properties concurrently during lead optimization, rather than sequentially improving single properties, has been a topic of growing focus in recent years [4]. In an idealized world (without cost or time constraints), all compounds from a hit series would be experimentally tested for potency, absorption, distribution, metabolism, excretion (ADME), and safety properties. With this full data package, hypotheses could be made about structure-property relationships for each property independently; decisions about which series to pursue and how to prosecute those series could then be made in an informed manner, taking into account the likelihood of successfully optimizing all the properties. A major factor to consider in prioritization would be the expected overlap between the hypothesized structure–property relationships. For example, structural changes that frequently improve stability to cytochrome P450 (CYP) metabolism (e.g., reduce lipophilicity) often work counter to other properties (e.g., potency).

From a practical standpoint, it is typically not feasible to obtain this complete idealized data package of all data. The number of data points required can be reduced by reducing the number of compounds profiled or by reducing the number of assays. From the viewpoint of trying to build structure–property relationship hypotheses, it is typically beneficial to profile a range of structures within a given series, so that the structural boundaries (within the extant chemical space) of "good" properties can be defined. At the hit triage stage, there is often greater value in profiling a range of structurally different compounds within a series than in profiling numerous highly similar compounds. The broader selection of compounds will hold the potential to generate more knowledge about the properties of the series. Methods for selecting these representative compounds can range from clustering using computational methods, to spanning a physical property range (e.g., lipophilicity), to manually selecting "diverse" chemical structures. The intent of selecting a range of compounds is to build a broader base of knowledge that will enable more accurate projections of the properties of future compounds in the series.

In terms of reducing the number of experimental assays employed, the use of computational models to predict specific in vitro endpoints has grown in recent years. In deciding how heavily to weight the predictions of a model, one must consider both the quality of the models predictions and the context in which those predictions will be used. It may be possible to use experimental measurements to confirm the predictions of a model for a given chemical series, and then to proceed with a better defined sense of confidence in the predictions of that model. There are useful examples of computational models for specific properties performing quite well; in these cases, model predictions may be useful not only for describing the properties of extant compounds, but also for virtual testing of proposed structure–property relationships.

At another level removed from experimental data are some general concepts that relate physicochemical properties to general trends in existing data. These concepts can be useful for evaluating the likely properties of hit compounds and series in the absence of much real data, essentially these can fill the gaps in an existing data set. They can also be very valuable for guiding medicinal chemistry thinking in the forward progression of a series of compounds. While real experimental data are clearly most valuable to evaluate the properties of extant compounds, the use of these general principles can be extremely powerful in selecting chemical strategies for the design of compounds that have yet to be made.

Figure 2 illustrates the key inputs to the hit triage decision making process. Data will come from multiple sources and different hits or hit series will likely have different quantities and types of data available. The critical project and medicinal chemistry decisions that must be made are defining the goals for each hit series and hypothesizing structural changes to reach those goals. The integration and balance of these data and decisions enables hit triage decisions.

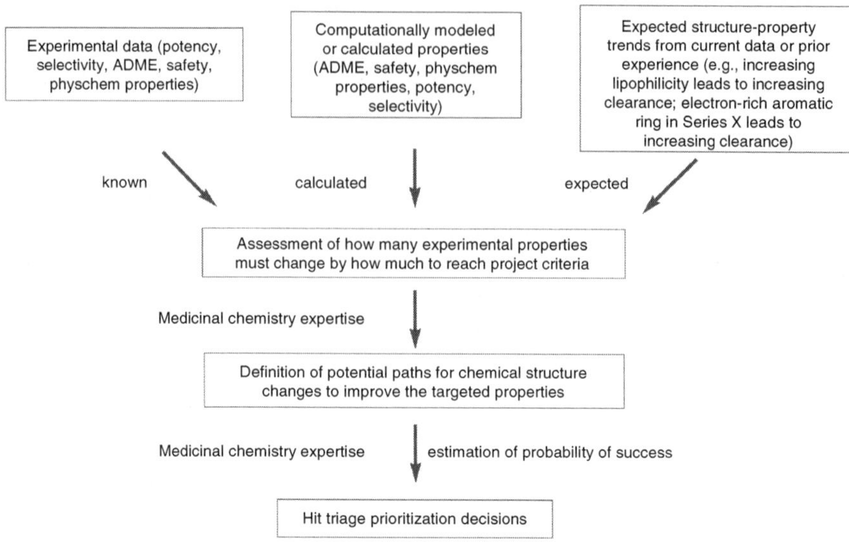

Fig. 2 Hit triage decision making requires inputs of experimental and calculated data and medicinal chemistry experience and expertise to arrive at an assessment of the probability of advancing individual chemical series

Keeping all of the above general considerations in mind, the next sections will cover sequentially various types of experimental data that can be examined in the course of a hit triage effort. As each is discussed, it is important to remember that a successful medicinal chemistry program is the ultimate goal. That success will almost always involve trade-offs between desirable properties, and those can be hard to define in a concrete fashion. Different projects may place a premium on one or more of these assays, given the particular needs of the specific therapeutic area, or the specific project. Simply put, there is no single way to do hit triage, and each effort has unique considerations. Still, some of the key consistently important parameters are discussed below.

2.1 Experimental Data: Potency

In almost all cases in drug discovery the search for a successful candidate molecule is the search for an appropriate ligand to bind to a molecular target. Since binding of a small molecule to a larger biomolecule is a physical chemical event, the binding of a ligand to its molecular target is driven completely by physical chemical forces [5]. These include Van der Waals (hydrophobic) interactions, hydrogen bonding, and ionic interactions. Although ligands are sometimes covalently attached to their targets by various reactions, these are relatively rare events and not often considered desirable interactions in a drug discovery setting [6].

Potency is an important property for a lead molecule (or series) when attempting to make decisions in hit-triage. At the early stages of a drug discovery program, the binding affinity of a compound is almost always considered as a surrogate for pharmacological efficacy. True efficacy, the induction of a specific biological response based on a molecular interaction, is the ultimate goal of a discovery team. For most of the last few decades, hard cut-offs were used that dramatically simplified the hit triage process. Teams would either work on the most potent hits, or would not work on any hit that was below some threshold or "dignity line" (1–5 μM, for example). When considering the hit triage process, the fundamental question to consider is how likely is a given series to produce a molecule with the desired potency to show efficacy. Understanding the assumptions that are built into that question (and all of their potential limitations) is a critical first step in a successful hit triage process.

In order to understand the potency of a given molecule, project scientists must first understand the nature of the measurement used to generate the potency value. If the molecule in question is from a literature report, there may be no in-house data to examine, and in that case the synthesis of the compound and the development of an assay usually becomes a critical first step to understanding the project landscape at that point. More commonly, however, the compound in question has been identified in some type of in-house screening process. These can take many forms, but they can broadly be grouped into two major categories:

1. Assays with a physical endpoint measurement
2. Assays with a biological endpoint measurement

2.1.1 Biophysical Assays

While biological assays are by far the most common, biophysical assays are steadily increasing in scope and popularity [7]. The rise in use of these assays is essentially paralleled by overall improvements in instrumentation, automation, and methodology over the last decade, which has provided multiple options for in-depth understanding of fundamental molecular interactions. These approaches almost always involve the direct measurement of binding of a molecule with a target of interest (protein, receptor, etc.). In this respect, these biophysical approaches differ significantly from assays that rely on a biological endpoint which will be discussed shortly. Since biophysical methods make direct measurements of molecular interactions, they can potentially reduce the need for complex validation assays [7].

In making these measurements, discovery teams can often gain key understanding of important physical parameters (binding rates, enthalpy, entropy, and even structural information). These can occasionally be lost in more traditional biological assays (particularly when run in a high throughput fashion). Also important to note – biophysical methods are often used in conjunction with fragment based approaches to drug discovery [8]. This is conceptually important, since the fragment approach identifies an overall smaller number of starting points compared to a typical HTS, and teams often have the time to invest in a more thorough understanding of each potential starting point. Additionally, since each hit is generally further from an eventual target profile, there is a critical need to do as much early work as possible to clearly define the scope of the problem at hand.

Without question, one of the most useful biophysical methods is co-crystallization and X-ray structure determination [9]. As with other techniques, automation and robotics are transforming the landscape of what was once thought possible. There have been published reports of parallel processing as many as 100,000 microcrystallization experiments in a single day [10]. The three dimensional binding information that is provided using this technique is particularly amenable to the central question in hit-triage: are there options for further improvements in these compounds? Of course, these improvements are not limited to enhancements in potency. X-ray structures can provide avenues for rational adjustment of a large number of properties by identifying areas of the molecules that may be amenable to substitution or chemical modification. There has been a significant amount of work recently in fragment approaches that are enabled by structure based drug design [11]. In many respects the fragment approach is ideally suited to projects which have X-ray crystal structures available. The fragments are small and relatively weak binders, but they often only possess one pharmacophoric element that binds to a specific feature on the target. If this interaction is identified by X-ray structure determination, then project teams can propose specific plans which maintain that critical interaction, and ideally optimize binding through other vectors in their fragments.

Even when co-crystallization is not feasible, if X-ray structures of the target are known, then there are opportunities to use computational techniques to try to understand the potential interactions of the compounds with their biological target [12]. This technique is inherently more limited when attempting to understand the binding of smaller fragments, as these are often small enough to bind in multiple potential conformations, and the calculated energy difference between these conformations can be extremely small, leading to relative uncertainty about the actual binding mode that is being accessed by a potential lead structure [13]. Even with the associated uncertainty, the low cost of virtual screening in comparison to screening of discrete compounds makes it a very attractive addition to the search for potential hits, especially for difficult targets.

Among other biophysical methods, NMR can also provide useful structural information. For an excellent recent review see Zartler et al. [14]. Critically important, NMR can often provide data to suggest that the binding of a novel molecular entity is causing the disruption of binding (displacement) of another inhibitor (or possibly the natural ligand). This information can be useful in building confidence in the hypothesized mode of binding and in understanding the effects of subsequent structural changes – e.g., if a modification is made, and as a result, the new compound no longer displaces the natural ligand of a given target – that change has fundamentally altered the binding mode of the new compound, relative to the starting analog. Traditionally, NMR screening approaches have required relatively large amounts of protein and of target compounds. This has resulted in the generation of NMR screening collections of ~1,000–10,000 compounds. Several groups have focused on the application of NMR techniques to the high throughput screening problem. Using $^{15}N/^{1}H$ correlation spectra, Hajduk and co-workers reported a strategy which would allow the screening of >200,000 compounds in less than one month [15]. Newer reports suggest that NMR can be used without labeling of protein or ligand in ligand competition binding experiments to screen extremely large numbers of compounds, and provide an estimated K_D of the ligand with a single point measurement [16].

More recently, developments in optical biosensors allow the precise measurement of ligand association and dissociation kinetics (k_{on} and k_{off}). Surface Plasmon Resonance (SPR), when used with immobilized proteins, can provide extremely sensitive measurements of the interactions of potential lead compounds [17]. The rate constants that are provided by this method can be critical to a clear understanding of a compound or series. For instance, a compound with a relatively low micromolar IC50 value in a biological assay may nonetheless display very slow dissociation kinetics, and thus, while being only a "weak" binder, could be an interesting starting point due to its unusual kinetic behavior. Overall, the use of this technology can build confidence in the binding behavior of a compound or series prior to the additional investment in optimization towards nomination of a clinical candidate [18]. Further, the knowledge of kinetic and thermodynamic parameters can help optimize both biological assays and decision-making in later stages of discovery [19].

As a complement to any (and perhaps all) of the above methods, calorimetry can be utilized in developing an understanding of the overall energetic behavior of the binding event [20]. The overall thermodynamics of any molecular interaction is the sum of both the enthalpic and entropic energy components of the species involved [21]. While these measurements have historically been somewhat limited due to a requirement for a significant amount of protein, new techniques have alleviated the situation substantially [22].

2.1.2 Biological Assays

For strictly biological assays, the assumption is often that the inherent binding affinity is a surrogate for pharmacological efficacy. Essentially, binding to a molecular target with higher affinity (driven by both enthalpic and entropic considerations) gives a higher likelihood of producing a desired biological effect. Most programs, especially at the hit triage stage, use this assumption as a given. The thermodynamics of binding is used almost exclusively to drive SAR development, while kinetics, due to more complicated and time consuming assays (or more expensive techniques, as discussed above) is largely not examined. Recently more and more reports of unusual kinetic behavior and the subsequent effects on the drug discovery process have been reported [23]. These reports have resulted in a growing awareness of developing an understanding of both kinetic and thermodynamic behavior of potential hits. Recent reports have described the integration of

HTS data and secondary screening for hit validation and determination of kinetic on/off rates. These data have been successfully incorporated into hit triage by enhancing the understanding of SAR differences between potential scaffolds [24]. In this example, those differences allowed informed decision making in the choice of which series to pursue and which to deprioritize.

Of course, traditional biological assays can provide direct functional data in the presence or absence of a test compound, i.e., is protein X inhibited in the presence of compound Y under a given set of conditions. Over-expression and purification, now of even historically difficult targets, have allowed discovery teams access to most potential drug targets. Expanding molecular biology techniques have facilitated generation of simple constructs for screening, and have enabled many screening programs that would have been difficult otherwise [25]. There are, of course, many caveats in setting up any given assay, and running under conditions which produce consistent data and biologically relevant results. The sheer number of permutations in ways to run screening assays and interpret screening data is far beyond the scope of this chapter. However, there have been several excellent reviews on the subject. In particular, the screening section of *Comprehensive Medicinal Chemistry II* gives an excellent starting point for the review of various methods and strategies for compound screening [26, 27]. Some assays are not a direct measure of inhibition or functional change, but rather rely on a series of interdependent biochemical events. The readout comes from a change in some downstream event in the presence or absence of a test compound. Often cell-based assays are done in this manner [28]. While these systems can be challenging to interpret due to complicated interactions and relationships within the system, they can also be used to provide more information. Specifically, because the compounds may be in a more biologically relevant environment, their effects on multiple endpoints can sometimes be monitored. In the best situations, the information that is gained in these studies can sometimes suggest new potential targets for therapeutic intervention.

2.2 Ligand Efficiency and Fragment Screening

While the high throughput screening approach has provided optimizable lead structures for many programs, it suffers from extreme inefficiency; compound collections at pharmaceutical companies now number in the range of 2×10^6. The cost of a typical full file screen can vary dramatically, but can easily approach (and in some cases exceed) \$1 million. Each HTS that is run is a significant undertaking and large commitment. Unfortunately, while that screening has provided some success, industry-wide it is fair to say that it has not delivered consistent success that has allowed teams to rapidly progress fundamentally new lead structures. At least part of this has been a focus on potency above other critical parameters, and less consideration of the ability to improve potency in smaller, more drug-like hits.

In the late 1990s, various reports which covered the energetics of ligand binding caused a re-examination of the mindset which had placed a premium on potency above other properties of a compound [29]. A new parameter was introduced which described the efficiency of binding, by dividing the binding energy of the compound by the number of heavy (non-hydrogen) atoms [30]. This was a clear theoretical break from previous hit triage approaches which centered heavily on potency as a dominant parameter to consider.

The "ligand efficiency" parameter (LE) allows the evaluation of the relative contribution to binding of all of the heavy atoms in a given molecule. Compounds which were potent binders, but high molecular weight would be less "efficient" (lower LE) than those that could accomplish similar binding with lower molecular weight. This parameter can be considered a key link between the traditional HTS approach and the fragment-based approach mentioned earlier, as it allows a meaningful and simple comparison of compounds (and series) of very different molecular mass. LE is a useful parameter for describing the interaction between a compound and its biological target. One must be careful to recognize that LE is not an end point in its own right, but rather one means of gauging the probability that potency and other parameters can be optimized concurrently. As a result, it is not appropriate to consider LE as an independent parameter; the context of the compound's other properties (absolute potency, lipophilicity, other experimental data) must also be taken into account. Similarly, the lipophilicity of a hit compound or series can provide a gauge of what that compound's properties are likely to be, as well as provide a direction for subsequent optimization of the series. A recent report from a group at Abbott lends significant support to this notion [31]. In this work, optimized compounds from 15 different projects were "deconstructed" back to their putative "fragment" starting points. The analysis showed that at each virtual step of the re-construction process, the overall binding efficiency index value remained constant for each target. This provides evidence for the concept that starting from a lead without appropriate efficiency will result in challenges in balancing potency and MW to achieve a final compound with desirable physical properties. It is natural, then, to consider the LE of the hits that are likely to be found in a normal high throughput screening effort. Traditional HTS approaches are often designed with bioassays that are optimized to identify single-digit micromolar binding compounds. When those methods are applied to screening a large collection of compounds, many of which have very high molecular weight (having been synthesized as part of optimization efforts in other projects), the results are generally predictable. Identification of an analog (or series of analogs) with molecular weights of ~450–500, and potency in the range of ~1–5 μM (i.e., low LE) presents a clear dilemma. Potency in this range can easily result from the sum total of multiple suboptimal interactions, making it difficult to replace any one structural feature and "jump" to truly efficient, high potency analogs.

Ultimately, when low LE series were chosen for further follow-up and medicinal chemistry optimization, modifications to the structure to produce more potent analogs would usually involve the addition of mass. This would move compounds to an undesired property space, and make the balancing of potency and PK properties difficult. Regardless, much of this work did occur, without regard to property space considerations, often in the name of high speed preparation of libraries of compounds to quickly build SAR within a series of interest. Removal of mass, while perhaps a desirable strategy to then allow a team to make progress towards PK objectives, would likely lower potency and compounds could go undetected in assays that were optimized for the desired high potency range. Thus, teams would often adopt complicated "cut and add" strategies where one would simultaneously remove portions of a potential lead structure and add functionality at other positions. Of course, successful analogs could result from these efforts, but they tended to be slow, erratic, and plagued by difficulty in making consistent comparisons with preceding compounds.

In choosing a fragment or series that has higher LE, teams could (to some extent) avoid the tradeoffs that were often encountered in the above approach. By definition, these hits would have high binding energy per heavy atom, thus every heavy atom was playing a more important part in the overall binding of the compound or series. Because of that, one could take a sensible approach to maintain that binding efficiency and perhaps build conservatively, adding specific functionality to these lower molecular weight compounds. As these smaller analogs grew, their potency could increase to more pharmacologically useful levels, and perhaps reach good potency levels without moving out of desirable property space. Medicinal chemists are now well aware of the Lipinski "Rule-of-Five" parameters [32, 33], which provide simple molecular descriptors that help to define high and low probability of success in overall absorption profile. As described above, building rationally from initially small fragments provides a chemistry team the opportunity to find analogs within these parameters. Conversely, an HTS approach may identify starting points which already fall outside of the desired physical property space, again leading to optimization plans that involve removal of functionality with re-introduction of new structural features. Overall, there is considerable and growing recognition that fragment approaches are a powerful tool for rapid advancement of drug discovery programs and in many cases can provide attractive alternatives, relative to traditional high throughput screening approaches [34].

2.3 Lead-Like vs Drug-Like Hits

Much of the above discussion of LE and good starting points can be distilled to the concept of "lead-like" vs "drug-like" hits. These terms were coined from consideration of the properties of hit compounds in conjunction with some general knowledge of how physical properties of compounds tend to change as a series is progressed [35, 36]. As described above, molecular weight and/or lipophilicity almost always increase as compounds are optimized for potency. Thus, drug candidates typically turn out to be higher MW and more lipophilic than the hits that were used as starting points [37].

The consideration of lead- or drug-like properties can be very useful in evaluating the attractiveness of hit compounds because of the implicit recognition that the properties of the optimized compounds are likely to be different [38, 39]. It is precisely the extrapolation of what will happen to the biological properties of a compound (or series) as its physicochemical properties change that is the central goal of hit triage. From a lead-like physicochemical properties starting point (MW <350, clogP <3, micromolar potency – perhaps a high LE, low molecular weight hit fragment hit), there may be significant extrapolation that must occur; from a drug-like starting point (e.g., MW 450, clogP 4, <0.1 μM potency – perhaps a kinase inhibitor from another advanced project, used as a lead in a new kinase project), there is much less extrapolation. Either of these types of hits may provide reasonable starting points for medicinal chemistry depending on the project context, but the way that their data are evaluated should be very different. On one end of the spectrum, a team may be less interested in the precise pharmacokinetic properties of a "lead-like" series, since they would understand that a significant evolution of structure and properties will likely take place before a development candidate could be identified. Conversely, a team will likely place a higher premium on having good quality experimental data (biology and pharmacokinetics) for a series that is more "drug-like," because those molecules will already share many of the characteristics of the eventual candidate.

3 Experimental Data: Pharmacokinetics – Absorption, Distribution, Metabolism, Excretion

A recurrent challenge in medicinal chemistry is identifying compounds that have an appropriate overlap between desirable potency and ADME properties. In deciding which series to pursue at the hit triage stage, one must both consider the properties of the hit compounds and make a probabilistic assessment of how the properties will evolve as additional compounds around the hits are synthesized. Experimental data are obviously the most effective means of identifying the properties of the hits; the assessment of how the properties will evolve is a judgment of the medicinal chemist. Factors that can contribute to this judgment include physicochemical properties, expected potency- or ADME-related SAR trends, predictions from computational models, as well as prior experience and intuition.

Because the goal of hit triage is to identify chemical series that hold promise for further optimization, an approach to characterize the ADME properties of a series, not just individual compounds is often useful. Where possible, characterizing the structure–ADME property relationship, in much the same way that a structure–potency relationship is defined, can be valuable for assessing the probability that a given structural series can be successfully optimized. The goals of this ADME property characterization are twofold: (1) to identify specific structural features that may be liabilities (benefits), and (2) to identify general structure–ADME property correlations.

These specific and general data inputs serve as the basis for making judgments on how substantially the structure of the hit compound(s) will need to change as the potency, ADME, and safety properties are concurrently optimized. These judgments form the basis of the lead optimization effort. As such, an important consideration is the selection of compounds to define the boundaries of the structure–ADME relationship of a series. In some cases, the compounds that are selected for defining the ADME relationships will not be the same set of compounds that have the most potent primary pharmacology.

The key question to assess with regard to pharmacokinetic properties is how likely optimized compounds will be to reach their intended target tissue, what concentration they reach in that tissue, and how long that tissue will be exposed to efficacious levels of the drug. Individual properties that are typically important to consider include permeability, solubility, clearance, volume of distribution, and distribution to specific tissues (e.g., penetration to the brain). The attributes that drive these properties can be divided into two general types: (1) specific structural features, such as particular groups that are metabolically labile, and (2) general physicochemical properties, such as ionization state or lipophilicity, that have a strong correlation to the end point of interest [40–42]. Understanding of these relationships for each series of compounds will significantly facilitate prioritization among series.

Because of the cost, difficulty, and low-throughput nature of collecting in vivo pharmacokinetic data, high throughput in vitro assays are commonly used to assess many of these ADME properties [43, 44]. Table 1 contains some of the most commonly used assays. A caveat with all in vitro assessments of pharmacokinetic properties is that the in vitro assays are set up as models of in vivo parameters. In most cases, validation of the in vitro results will not occur until in vivo studies (in humans) are executed, and the uncertainty that is associated with these assays as models needs to be considered. As part of the effort to characterize series, in vivo PK experiments at an early stage (perhaps as early as hit triage) may be considered, particularly for "drug-like" hits. Due to the cost of in vivo experiments, it is not common to run them in the hit triage phase. Ideally, when they can be run, the main objective of these experiments is often to validate (or build) a correlation between

Table 1 Common in vitro assays to assess ADME properties of hit compounds

Clearance
Human (rat) liver microsomes
Cryo-preserved or cultured human (rat) hepatocytes
Recombinant CYP450s (also for safety assessment of drug–drug interactions)
Permeability/absorption
Caco-2 cells
MDCK cells
PAMPA (artificial membrane)
Solubility
Kinetic solubility (from DMSO solution)
Thermodynamic solubility (from crystalline material)

in vitro and in vivo assays. As with in vitro assays it is important to also recall that preclinical in vivo studies are only models for human pharmacokinetic outcomes.

3.1 Clearance

Clearance is a critical parameter because of its role in determining a drug's dose size and frequency. First-pass clearance in combination with absorption determines a compound's bioavailability. Clearance and absorption in combination with potency determine dose size. Clearance and volume of distribution determine half-life, and thus dosing frequency.

The assessment of clearance is complicated by the numerous mechanisms by which compounds may be cleared from the body. These mechanisms include oxidative metabolism, most commonly by CYP enzymes, but also in some cases by other enzymes including but not limited to monoamine oxidases (MAO), flavin-containing monooxygenases (FMO), and aldehyde oxidase [45, 46]. Non-oxidative metabolism such as conjugation or hydrolysis may be effected by enzymes such as glucuronyl transferases (UGT), glutathione transferases (GST), amidases, esterases, or ketone reductases, as well as other enzymes [47, 48]. In addition to metabolic pathways, parent compound may be excreted directly via passive or active transport processes, most commonly into the urine or bile.

From an early drug discovery and hit triage perspective, a simplification that is typically made is to focus initially on CYP metabolism that occurs in the liver. This simplification is appropriate for the majority of compounds – some compounds will be cleared by other mechanisms, but it is generally assumed that they must be relatively stable to CYP metabolism to achieve a desirable PK profile.

With this focus on CYP and liver metabolism, most companies have established high throughput assays to measure compound stability in the presence of human (or preclinical species) liver microsomes [49]. Disappearance of starting compound from an incubation with microsomes is monitored. Measurement at a single time point enables a rank-ordering of compounds for stability based on percent of parent compound remaining; acquisition of data at multiple time points allows determination of half-life, intrinsic clearance, and extrapolation to a predicted in vivo clearance [50].

Other assays for assessing CYP clearance are also employed, although often less widely or with lower compound throughput. Recombinant CYP enzymes allow the determination of the kinetic parameters for metabolism of individual compounds by individual CYPs. Recombinant CYPs also provide an avenue to assessing and understanding the potential for drug–drug interactions that may occur between two or more compounds.

Hepatocytes, whether freshly cultured or cryo-preserved, can provide an assessment of not only CYP metabolism but also clearance by other metabolizing enzymes and potentially the role of transporters [51]. The accuracy of the data is of course dependent on how well the proteins in the hepatocytes function after culturing or freezing.

In the interpretation of data from a metabolism assay, care must be taken to consider the role of nonspecific protein binding in the assay [52]. Within a given chemical series, protein binding typically increases with increasing lipophilicity. The importance of protein binding on predicting in vivo clearance from in vitro measurements is well described. At the hit triage stage, when compounds from multiple chemical series of potentially different lipophilicities are being considered, the differences in lipophilicity should be taken into account if clearance data are not corrected for protein binding; apparent metabolic stability in lipophilic compounds may be an artifact of high protein binding.

In evaluating the clearance properties of a series, identification of a correlation between lipophilicity and clearance can provide a useful general framework for the design of subsequent compounds. The absence of a general correlation between decreased lipophilicity and reduced clearance (i.e., compounds with low logD are rapidly metabolized) may be an indicator of a specific structural feature that is a metabolic liability, which can form the basis for specific structural hypotheses to be tested. General structural features that are often thought to contribute to CYP metabolism include high lipophilicity, a high number of rotatable bonds, and electron-rich aromatic rings [53]. Although most chemical series do show a general correlation between lipophilicity and CYP-mediated clearance, there can be significant differences between series. These differences may be an important decision-making factor in prioritizing series during hit triage.

The figures that follow provide examples of some ways in which in vitro clearance data for two series can be compared and assessed to identify key questions, trends, or hypotheses. While the data presented here are for clearance in a human liver microsomal (HLM) incubation, the analysis could be applied in the same way to other data sets – including other experimental ADME or safety end points, or computationally predicted end points.

Figure 3 represents two series of compounds where a significant amount of experimental in vitro clearance data is available. In looking for general trends, neither series appears to demonstrate a robust correlation between clearance and lipophilicity. Series A (open squares) series shows a rough trend toward greater probability of having stable compounds at low (clogD <2) or high (clogD >4) lipophilicity. For Series B (black triangles) in particular, a key question is whether there are specific structural features common to the compounds that may be metabolic liabilities – the few more stable compounds will be valuable points of comparison.

Figure 4 shows the same compounds as Fig. 3; the clearance data are corrected for microsomal protein binding as calculated by a computational model. Experimental measurements for a representative subset of these compounds confirmed that the model was reasonably accurate. Series A demonstrates a strong correlation between free clearance and lipophilicity; depending on the level of clearance desired, clogD could be used as a guideline for the design and selection of subsequent compounds. Assuming a desired Clint,free of <100, clogD of $\sim$1–3 might be the target range. A critical question in deciding the relative prioritization of this series would be how this target lipophilicity range for clearance overlaps

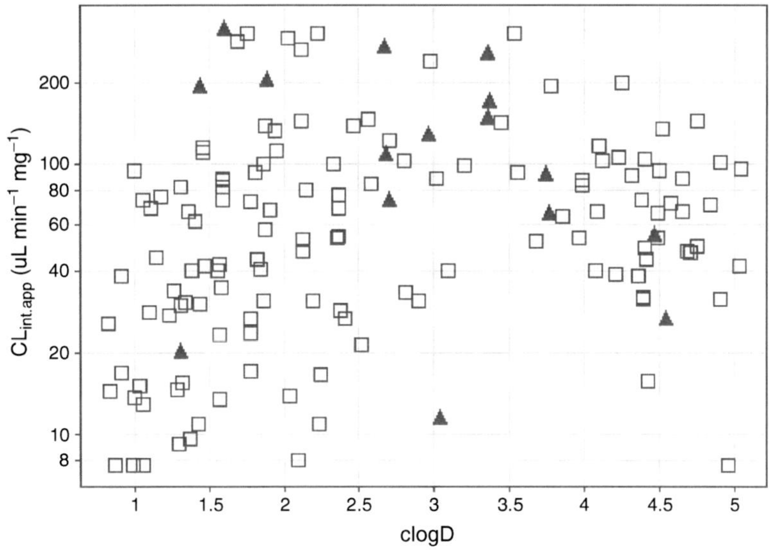

Fig. 3 Human liver microsome apparent intrinsic clearance (Clint,app) vs clogD. *Open squares* and *filled triangles* represent two different chemical series (series A and B, respectively)

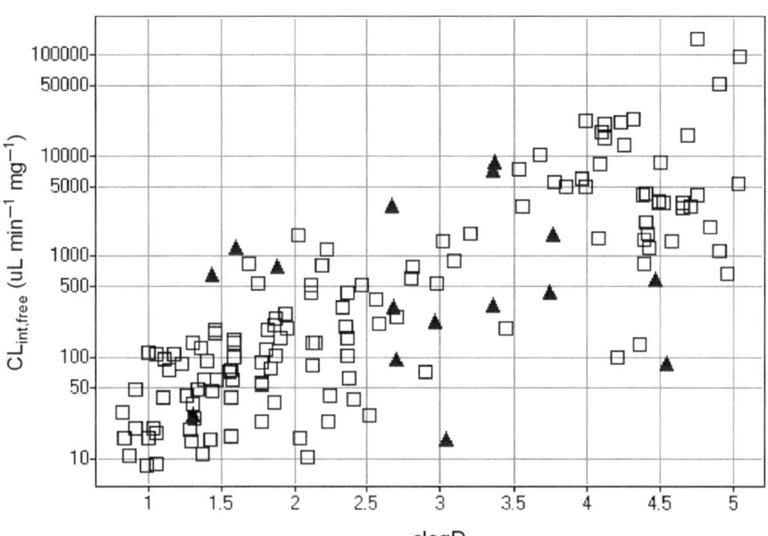

Fig. 4 HLM Clint, free vs clogD. HLM Clint, app corrected for microsomal protein binding using a computational model for microsomal binding. *Open squares* and *filled triangles* represent the same chemical series as in Fig. 3 (series A and B, respectively)

with the desirable lipophilicity ranges for other properties (potency, safety, etc.). For Series B, there is still not a strong correlation between clearance and lipophilicity. For both series, a more detailed analysis of the structure–clearance relationship may

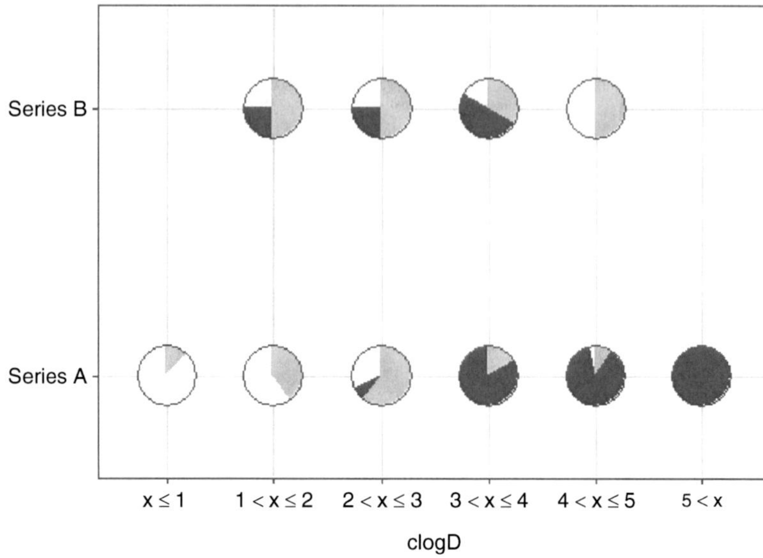

Fig. 5 Pie charts for the two chemical series, colored by binned HLM Clint, free (*white* – low; *grey* – moderate; *black* – high) and arranged by binned clogD. Series A (*bottom*), series B (*top*)

identify specific structural features of importance; in particular, "outlier" data points (e.g., compounds that are unusually stable at a given clogD) often contain information that can lead to a new direction for the evolution of structure–property relationships.

Figure 5 shows an alternative view of the free clearance data for these two series. Free clearance values were binned into three groups, and the compounds in each series were binned by their clogD values. By the progression in the pie charts, the correlation (or lack thereof) between clearance and lipophilicity is evident. This type of view does not directly highlight the individual "outlier" compounds that may contain extremely informative structural information, but it is valuable for identifying general trends.

On the assumption that properties other than clearance were similar for Series A and B, specific structure–clearance relationships and the general clearance-lipophilicity trend for Series A are important factors for hit triage prioritization. If the specific structure–clearance relationships do not differentiate the series, Series A might be favored for more having a more straightforward strategy to optimize clearance (target clogD <3). It should be emphasized that this strategy is a probabilistic one – the identification of low clearance compounds is more likely at lower clogD; there may well be some higher clogD compounds that would be stable, and not every lower clogD compound will have the desired clearance profile.

Figure 6 is another view of the data for only the compounds in Series A. Compounds were binned into groups based on both their clogD and MW.

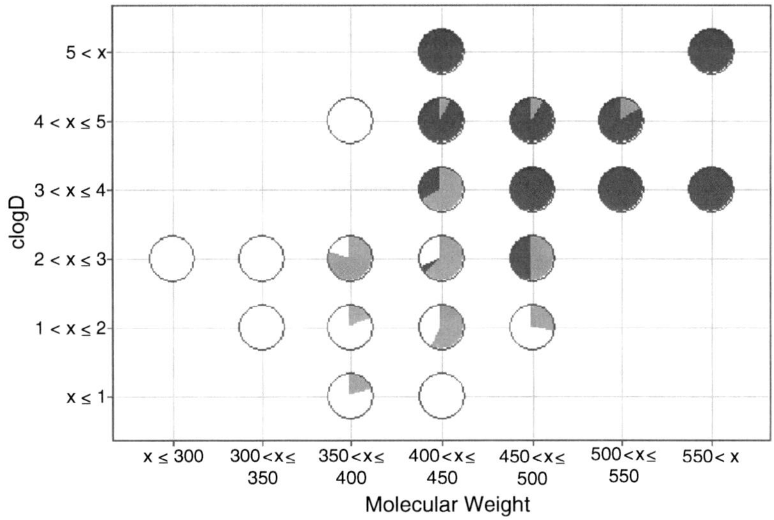

Fig. 6 Pie charts for series A only, colored by binned HLM Clint, free (*white* – low; *grey* – moderate; *black* – high) and arranged by binned clogD and binned MW

The trend toward decreased stability of compounds with increased lipophilicity and MW is obvious. This analysis can provide information about the roles of these two properties in contributing to clearance for this series. Again, the important question for hit triage decision making is how the hypothesized structure–clearance trends correlate to the property trends for other relevant biological properties.

3.2 Permeability/Absorption

The permeability of compounds across biological membranes is important for compounds to reach their biological targets. Many drug discovery programs target oral dosing and absorption of compound in the gut; thus permeability across these membranes is an important parameter to understand. A high percentage of drug targets have an intracellular location, requiring compounds to cross cell membranes in the target tissue; an additional subset of compounds must cross the blood–brain barrier to reach their target tissue. Due to the overall importance of maintaining good permeability in most traditional programs, many teams will want an early understanding of the likelihood of permeability problems with the series of interest.

Typical early in vitro permeability assessments measure the rate of flux of a compound from one side of a barrier to another [54, 55]. The barrier has historically been derived from a cell line, most commonly Caco-2 or Madin–Darby canine kidney (MDCK) cells. In the last several years, there has been substantial work and significant progress in the development of parallel artificial membrane permeability assays (PAMPA) [56–58]. PAMPA assays can decrease the cost of permeability assessment as compared to cell-based assays. PAMPA also provides the opportunity to vary the membrane and donor or acceptor buffers, which can allow an empirically-derived modeling of different biological membranes.

PAMPA is typically used to make a prediction of the passive, transcellular absorption of a compound. Compounds which may be absorbed by a paracellular mechanism or may be substrates for active transport (uptake or efflux) are usually better assessed in a cell based system. A combination of assays can be applied to gain a greater understanding of the permeability and transport properties of a compound.

The study of active transport mechanisms has grown substantially in recent years, with transport proteins such as P-gp, BCRP, and MRP-2 among the most studied [59]. Several types of in vitro assays to assess substrates of transporters have been established; these include assays directed toward intestinal and biliary efflux [60]. Assays that measure passive and active transport are also used to assess penetration of the blood–brain barrier. In addition to the assays described above, transfected cell lines that overexpress transporters present in the blood–brain barrier are also employed [61].

In vitro assessments of permeability are commonly used to bin compounds as high, low, or moderate permeability. At the hit triage stage, compound series that demonstrate overall low permeability may be a cause for concern (assuming high permeability is desired), particularly if there are not obvious structural features that might cause poor permeability. Considering that crossing a membrane requires a compound to pass from an aqueous environment to a lipid and back to water, structural properties that can often lead to poor passive permeability include the extremes of polarity or lipophilicity (e.g., high ionization or high hydrophobicity) or high hydrogen bonding capacity [62]. Structural rigidity and compound shape are also likely to play a role in determining a compound's permeability properties. Lipinski's "Rule-of-Five" was a landmark proposal for identifying a subset of physicochemical properties that tend to correlate with poor absorption.

Figure 7 shows the relationship between permeability in a MDCK cell assay and clogD for two series. Series C (filled squares) demonstrates good permeability, although only a relatively narrow clogD range was characterized. The importance of characterizing permeability for this series at lower clogD before making hit triage prioritization decisions will depend on whether optimization of other properties are likely to drive future compounds toward lower clogD. Series D (open circles) shows variable permeability across a wider clogD range. A more detailed structural analysis will be valuable to assess the impact of specific structural features on the permeability properties of this series.

Figure 8 shows PAMPA data for a subset of compounds from the two series in Fig. 7. The potential concern about low permeability for Series D is confirmed between the MDCK and PAMPA data. Series C, with the exception of one compound, also demonstrates good correlation between MDCK and PAMPA permeability; the compound with low PAMPA permeability should be further analyzed for relevant structure-permeability information.

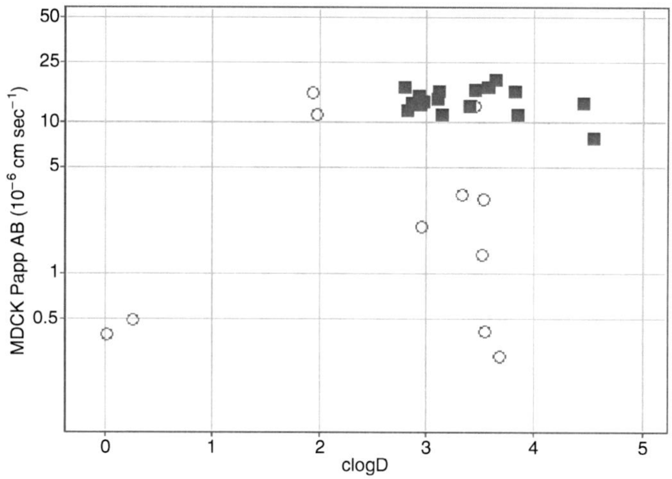

Fig. 7 MDCK A to B apparent permeability vs clogD. *Filled squares* and *open triangles* represent two different chemical series (series C and D, respectively)

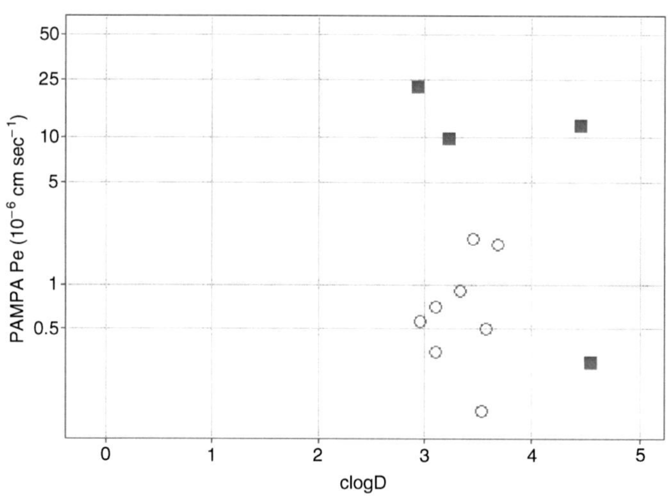

Fig. 8 PAMPA flux vs clogD. *Filled squares* and *open triangles* represent the same chemical series as in Fig. 7 (series C and D, respectively)

3.3 Solubility

Compound solubility is important because it affects the bioavailability of compounds in vivo, the behavior of compounds in in vitro assays, and the ease with which preclinical in vivo studies can be run. Solubility assays that measure either kinetic or thermodynamic solubility are generally employed in drug discovery [63]. Kinetic solubility assays have been developed to increase compound throughput, and are thus often used during hit triage and early in chemical series progression. Thermodynamic solubility assays are generally more resource-intensive and are usually employed as compounds move closer to being candidates for drug development.

Kinetic solubility assays generally utilize DMSO-stock solutions of test compounds and measure the extent to which compound comes out of solution upon dilution with an aqueous solution. Assay times can be short (seconds to minutes), with detection by light scattering or by physical separation of precipitate and quantization of remaining solute, thus enabling higher throughput screening. Thermodynamic solubility assays generally start with solid compound sample (ideally crystalline) and measure solute concentration after a time period of hours to days. Various studies have examined the correlation between kinetic and thermodynamic solubility measurements, without consistent conclusions [63]. The correlation between measurements may vary by chemical series or by physicochemical property class. A general concern about kinetic solubility assays is that they will overestimate thermodynamic compound solubility due to the presence of small amounts of DMSO or supersaturation of the test solution. Recent research has been directed toward the development of high throughput assays that more closely mimic thermodynamic solubility measurements [63], for example by evaporation of organic solvent from the test sample before solubility analysis.

At the hit triage stage, it is most common to be able to characterize sets of compounds in a kinetic solubility assay. In the assessment and utilization of these data, the potential disconnects between kinetic and thermodynamic solubility must be considered. Low kinetic solubility for a series of compounds should lead a project team to be concerned about the behavior of compounds in biological assays and buffers, as well as the potential for optimizing drug-like properties in that series. Conversely, while high kinetic solubility is a desirable property, chemists should still remain cognizant of the need to assess thermodynamic solubility as compounds are further optimized.

The solubility properties of a compound are determined by the fundamental interactions of the compound in the solid state and by its interactions with aqueous solution [64]. Chemical properties that can be manipulated to impact solubility include ionization state, lipophilicity, hydrogen bonding capacity, and the three-dimensional shape of compounds.

4 Experimental Data: Safety

Just as there is a significant challenge to understand the relationship between expected potency and expected PK parameters in order to define the probability of identifying a low dose compound, there is a similar interplay between potency and safety endpoints to identify nontoxic compounds. Toxicity represents a significant challenge in the development of compounds, accounting for as much as 30% of early attrition in development [65]. The cost associated with this attrition is substantial, and has caused pharmaceutical companies to attempt to solve these problems at earlier stages of the discovery process. Thus, hit-triage teams are now challenged with the problem of attempting to extrapolate early (and sometimes incomplete) data on sub-optimal compounds, to identify reasonably high probability areas on which to focus. As with other properties, experimental data and judgment are key to solving this problem. In some cases, that data and judgment can be augmented by predictions from computational models. In the subsections that follow some of the most significant safety assays are highlighted, and their contributions to decision-making are discussed.

4.1 hERG

In assessing the safety risk of a given molecule (or series) a key question to ask is how likely is this analog to encounter interaction with ion channels. In particular, a key cardiac ion channel: the human ether-a-go-go related gene (hERG) can present a serious safety consideration for teams assessing the likelihood of finding compounds that are safe as pharmaceutical agents [66]. Interaction with this ion channel can cause prolongation of the QT interval and potentially lead to the development of severe cardiac arrhythmia known as "torsades des pointes" [67]. Since this condition can cause sudden death, it has produced a massive industry-wide effort to understand and avoid the specific molecular interactions involved.

A number of approaches have been undertaken to unravel the specifics of the hERG interaction of pharmacologically relevant compounds. While screening for hERG activity itself is possible, the industry standard assay is a patch clamp approach to measure functional activity [68, 69]. This assay does not support high throughput screening of large collections of compounds. While there have been reports of higher throughput [70], assays of this type are still more "medium throughput" and would not allow the efficient measurement of potential hERG liability of a collection of thousands of compounds. In order to circumvent this limitation, groups have developed radiolabeled competitive binding assays which give an indirect view of potential hERG channel interaction and thus cardiac risk [71]. More recently, fluorescent dofetilide analogs have been reported that allow a very high throughput fluorescence polarization binding assay [72]. This type of assay could dramatically reduce the reliance on radiolabeled substrate, and have benefits for reduced radioactive waste disposal and lower overall environmental impact.

Attempts to build predictive models based on common pharmacophoric elements have had some modest success. In the most general sense, a basic nitrogen atom which is substituted by aromatic or otherwise hydrophobic groups is a clearly problematic motif [73, 74]. However, there are many compounds which interact with hERG which do not contain these features, and newer pharmacophore models have been proposed to account for these compounds as well [75]. To complement pharmacophore models, descriptor-based computational models have also been published and used to make progress, generally within defined series [76, 77]. Many efforts have also examined the potential correlations with physical properties of the compounds, and high lipophilicity is strongly correlated with risk of hERG activity [78].

From a hit-triage perspective, this presents a complicated dilemma. For any given series, there may be a structure–activity relationship, or physical chemical property information which can lead a team away from an hERG problem. However, it is rarely the case that a wealth of that information exists when a team must decide on a potential series for follow-up. More commonly, a team may have hERG binding data (or a surrogate such as dofetilide assay data) on a small number of compounds (or just one compound). Those data, then, must be evaluated in context with the rest of the compounds' properties, in order to make a reasonable assessment of the impact of the hERG activity. A final complexity involves the relationship between binding affinity for the hERG channel, and binding affinity for the target of interest. The therapeutic window that is relevant is a difficult question to assess when dealing with incomplete data sets and early data for a series that is not yet optimized in either direction: for potency, or away from toxicity endpoints such as hERG. Somehow, assessment of the relative likelihood of this type of optimization is the critical piece of information to consider for each series that presents itself at the hit triage stage. This is always case dependent. If series A, for example, shows significant hERG liability for all tested compounds, irrespective of chemical structure changes, physical property changes, or potency variations, then it could confidently be deprioritized. A different series, B for instance, might possess significant hERG liability, but chemistry follow up might be more attractive if

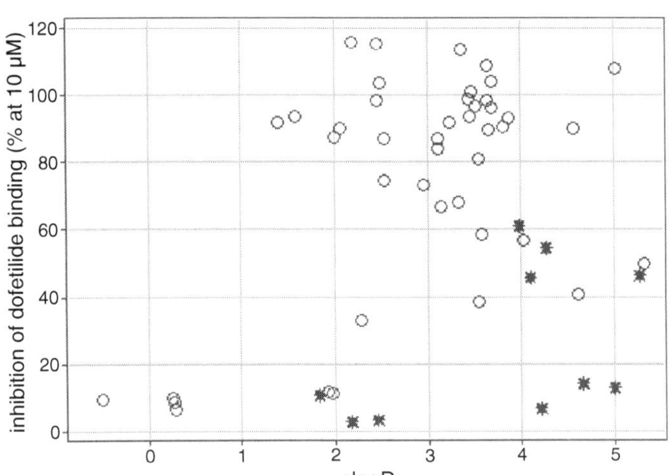

Fig. 9 Inhibition of dofetilide binding (percent inhibition @ 10 μM) vs clogD. Series A (*black stars*) and series B (*open circles*) are structurally distinct

that liability disappears at a lower logP range, or in the absence of a specific structural feature that is not required for potency.

Figure 9 provides an example. Two different series (A and B) are plotted. These series each have unique structural features, and these result in differential behavior in a high-throughput dofetilide screen. While series A contains three compounds that have some significant inhibition of 45–60%, those only occur at a logD of 4 or greater. Series B (open circles) does contain a few scattered compounds that have very little inhibition, but those only occur at relatively low logD. The vast majority of the compounds in series B have significant dofetilide inhibition across a large range of logD values. This indicates a much more severe dofetilide problem in series B, which cannot be solved by simple lipophilicity modulation. These data can help to inform a choice between the two series in hit-triage.

4.2 Genetic Toxicity

While the precise mechanisms of toxicity are not always well understood, the risks associated with the introduction of a xenobiotic into a living organism can be substantial. Among the most serious of these interactions are those that result in genetic damage which can ultimately lead to carcinogenesis. While cancer can arise from many specific mechanisms, compounds can be considered genotoxic carcinogens if they interact directly with DNA, or with any of the processes that replicate and repair DNA. The entire pharmaceutical industry has spent significant time and effort on the problem of reducing the incidence of genetic toxicity in the nomination of compounds for clinical development.

Because of the obvious need to avoid the possibility of inducing cancers in patients or healthy volunteers, regulatory agencies require that there be evidence that new clinical test compounds are safe and unlikely to cause genetic damage. Currently, these requirements include a safe result in a bacterial gene mutation (Ames) assay, and an assay for chromosomal damage in a eukaryotic (mammalian) cell. Prior to entering Phase 2 clinical development, tests for chromosome damage in rodent bone marrow are required. Most often, all of these are conducted prior to beginning Phase 1 trials as a way to lower potential risks to volunteers, and as a way to lower costs in the event of a positive finding of toxicity.

The Ames test involves the reversion from a his− to his+ phenotype in any one of multiple bacterial strains (usually five strains are tested simultaneously). If the addition of test compound to a his− strain of bacteria allows them to grow on histidine deficient media, the obvious conclusion is compound-induced mutagenesis and a high potential hazard for the compound being carcinogenic. This test can also be conducted in the presence or absence of metabolic activation, in order to provide more information on potential risks (i.e., the parent compound may not be mutagenic, but the primary metabolite may present a safety risk). In practice, a positive Ames test almost always leads to discontinuing work on a compound of interest, and so these data are always collected prior to nomination of a compound for development.

In conjunction with Ames, there are other in vitro tests that are important predictors of potential genetic damage. The most common of these is the in vitro micronucleus assay. This assay gives a readout on compound induced chromosomal damage by studying the formation of micronuclei, chromosomal material that is not incorporated into the nucleus during cell division. This involves a manual assessment of cells that have been exposed to test compounds (both with and without metabolic activation, as in the Ames test). With the proper staining techniques, micronuclei can be identified, counted, and characterized as whole or fragmented parts of DNA. Due to the nature of the analysis involved, this is a low throughput assay, and it requires a relatively large amount of test compound (~30 mg).

The difficulty in understanding the scope and relevance of genetic toxicity in the hit triage process is the same as the difficulty in understanding other parameters: researchers are often asked to make a judgment about the relevance and significance of a risk that may be identified in an early compound, or perhaps in a structurally related analog. Since these studies require longer times and larger amounts of compounds, they may be difficult to do on multiple analogs, and impossible to do in a high throughput fashion to provide large amounts of data on a series. Thus, it is difficult to judge how prevalent these effects might be. Practically speaking, there are often insufficient data to judge accurately the scope of the problem.

Several solutions to increase the quantity of available data have been published recently, including a microplate Ames assay (Vitotox) that is very fast (a single afternoon), requires less than 1 mg quantities of compounds, and detects genetic damage using a light emission output assay. This can be run in 96-well format, and the results in one study were ~94% concordant with the traditional Ames test [79]. A similar approach to the micronucleus assay has resulted in the yeast DEL screen, an assay which can detect DNA deletions in *Saccharomyces cerevisiae* [80]. While this approach has reasonable concordance with the in vitro micronucleus assay (~70–80%), there are potential issues with differential compound permeability in the yeast cell wall relative to mammalian cell membranes. Another approach involves automation of the scoring of the traditional micronucleus assay, including a recent report using flow cytometry [81].

Also applicable is the development of in silico models to flag potential compounds (or series) as risks for genetic toxicity. Like any model, these would be far cheaper, faster, and have no physical sample requirements. The development of these models has likely been slowed by the lack of freely available toxicology data; however, certain models are applicable and widely used (DEREK for Windows, HazardExpert and ToxBox) [82] and new models are frequently discussed in the literature [83–89]. These models should continue to improve in conjunction with the increased data available from the improved throughput in modified genotoxicity assays. As with any other models, their use as early screens (to indicate potential risks) can help focus limited resources, and potentially help shorten timelines, by upfront identification of problematic compounds. These early indications can be followed by more significant investment in Ames or micronucleus screening, which can help to put the extent of the problem in perspective.

Finally, the decision of whether to proceed with medicinal chemistry optimization, then, may be based on several strategies. In the first, an attempt is made to avoid a risk completely. This is a possibility when options exist (i.e., pick between two series for the one with the best profile). The other approach is an attempt to mitigate a risk (i.e., decide to work on a series with a potential issue). This approach can be viable when other options are unavailable or in some cases when the severity of the disease warrants the risk (such as cancer). This is always done with the understanding that more effort will be required to monitor continually safety endpoints than in a traditional program. As in any hit-triage decision, the specific genotoxicity data will need to be weighed against everything that is known about a particular compound (or series), and an overall assessment of the profile will need to occur. The severity of the outcome with this particular endpoint, however, means that for most practical purposes (and non-life threatening indications), it would be rare to work on series that had problems when any other options exist.

4.3 Reactive Metabolite Formation, Mechanism-Based CYP Inhibition, and Relationship to Toxicity

Another component of toxicity of a given compound or series is the ability to cause adverse drug reactions. There has been a significant accumulation of evidence over many decades which provides a clear framework for understanding certain structural liabilities in compounds [90]. While the purpose of the CYP enzymes is to modify xenobiotics to allow for clearance to occur, in some specific cases this metabolism can lead to reactive (electrophilic) species which can then go on to react with biomolecules. These reactions can be harmless, or can lead to a variety of increasingly severe outcomes: rash, mild to severe autoimmune responses, hepatotoxicity, genotoxicity, and even death. There is a compelling accumulation of evidence suggesting a causative role of the reactive metabolite formation in adverse drug reactions; however, most of this evidence remains circumstantial. Recent reviews have covered the relationship between bioactivation of compounds and possible links with toxicity [91–94].

One specific type of reaction which can occur with these reactive intermediates is reversible or irreversible inhibition of CYP enzymes. When this occurs irreversibly, through covalent modification, this is termed “mechanism-based inactivation” (MBI). An excellent review of this phenomenon has recently been published [95]. Beyond hapten formation and autoimmune response, there is an additional risk from these events. The slow turnover of the inactivated CYP can lead to drug–drug interactions, potentially exposing patients to unanticipated high levels of the drug in question, or other drug substances that they may be taking, leading to possible overdose and associated toxicity. Detailed mechanistic studies can help describe the level of risk associated with these drug–drug interactions, but those are very unlikely to be undertaken in an early-stage of drug discovery. As a result, most companies have compiled both in-house and literature data on specific structural features that are likely to predispose a compound to this problem. Many examples are described by Kalgutkar et al. [95], and have also been described explicitly in an excellent review by Blagg [96].

In order to provide some experimental context to the possibility of reactive metabolite formation, assays have been developed that detect the formation of glutathione adducts post-incubation with human liver microsomes [97]. Of course, many caveats exist with such assays, such as detection sensitivity, quantization, relevance to in vivo situations when other clearance pathways are competitive, trapping with non-glutathione biomolecules, and overall relevance to adverse events and toxicity. However, this type of early screening can function as a reasonably efficient means of identifying a risk which can allow some analysis of options at the hit-triage stage. The development of increasingly higher-throughput versions of these screens will allow even more compounds to be analyzed [98], and provide bigger data sets to build computational models.

In terms of decision-making in hit triage, the options are quite similar to other toxicological endpoints. The full extent or relevance of an early finding to the profile of the eventual clinical candidate molecule might be very difficult to predict. As with other toxicity, the specific indication is a relevant consideration. A theoretical risk of toxicity is less of a concern in a life-saving therapy for cancer than in a chronic treatment for asthma. However, as an operational consideration, if there is a choice between multiple series where one appears less likely to form reactive metabolites, it would be reasonable to minimize risk. An alternative approach would be to try to correlate the reactive metabolite formation to a specific structural feature, using the large amount of literature available to make these connections. When these problematic structural features are required for pharmacological activity, weighing potency and possible toxicity is more difficult. Relying on in vitro and in vivo screening can provide a path forward, but this screening approach always adds complexity, cost, and delay in later stage optimization efforts. Even worse, the ultimate outcome of these efforts, and their relevance to patients will not necessarily be known for a given project until late in the development process with a single compound, perhaps as late as Phase 3 trials, when an enormous investment in time and money has already been made.

4.4 Broad Ligand Profile Screening

A concern with any molecule is that it selectively interacts with the target (or targets) of interest. It is quite common for projects to set up counterscreens in order to profile molecules vs related targets (isoforms that might have undesirable biological activity, other proteins from the same family with high sequence homology, etc.). However, one general concern is the behavior of a given compound or series when profiled against large panels of enzymes and receptors, each of which may have its own pharmacological relevance. In the ideal situation, teams would identify compounds that act at single targets (or a well defined small group of targets) and have no significant off-target activity; in reality, this is rarely the case.

For many years it has been routine practice to gather data on potential clinical candidates, and to try to understand relevant off-target pharmacology. This has resulted in a significant accumulation of internal data at pharmaceutical and biotech companies. One way to accomplish this is to determine percent inhibition or binding at a single high concentration (1 or 10 μM). This can be followed, of course, with full dose-response curves in assays where there is significant interaction. There are multiple vendors that provide screening in various panels, allowing teams to chose panels with targets of interest (kinases, receptors, etc.), and perhaps to collect data from multiple sources.

As with any other potential toxicity endpoint, there is a push to collect relevant data earlier in the discovery process, and in fact this can occasionally become part of the hit triage process. In practice, the expense of collecting these data for a significant number of compounds often prevents broad ligand profiling from being used as a screening tool. However, there have been reports of using broad ligand screening as a tool in moving from a hit to a lead and making decisions in the hit triage stage. Poulain et al. describe their findings that compounds that are structurally similar to promiscuous inhibitors have a higher likelihood of being promiscuous themselves [99]. This finding is described generally as insight into the structure profile relationships, and the concept has been extended to describe a compound's specific biological spectrum [100]. Fliri et al. argue that a compound's specific pattern of interaction with biological targets of interest generates a profile that can quantify similarities and differences between molecules, and provide a rational basis for quantifying the effects of structural changes between analogs. Regardless of how the data are used and interpreted in specific instances, the ability to assess potential interactions earlier in the drug discovery process should improve decision making and allow teams to focus on potential risks at earlier time points.

4.5 Computational Models

In a hit triage decision making process that blends the use of experimental data with expected general property trends and principles, there are situations where it is not feasible to obtain sufficient data to identify experimentally property trends for ADME or safety endpoints (either due to a small number of hit compounds in a series, or due to limited experimental capacity). Computational models for these parameters may provide some useful information when integrated with other known information [101].

There has been extensive work on computational modeling of ADME and safety properties in recent years, but the field is still evolving [102, 103]. There are two key limitations on the use of models for these endpoints. One limitation is technical – the quality and accuracy of the models for the chemical space of interest [104, 105].

The other limitation is intellectual – the appropriate interpretation of model predictions and integration into the decision-making process [106, 107].

On the technical side, many different model building techniques are being explored and utilized. A fundamental constraint on the application of any model is the quality and availability of the data that it is built upon. In drug discovery, where the true data of interest (human in vivo parameters) are difficult to obtain and scarce, in vitro or preclinical in vivo experimental models are used to generate larger data sets and to guide decision-making. Most commonly, computational models are then used to predict these in vitro or preclinical endpoints.

The properties that have received the greatest attention with regard to computational modeling are solubility, intestinal absorption (passive and active), CYP metabolism and drug–drug interactions, brain penetration, and hERG binding, although almost all ADME and safety properties that have adequate data sets have been modeled [108-116].

Key challenges in utilizing model predictions include gauging the likelihood that the current chemical matter of interest falls within chemical space that is reasonably predicted by the model and translating model predictions into decisions on what types of compounds to pursue next. For gauging the relevance of model predictions to current chemical matter, the use of selected experimental data points can be of significant value. In utilizing model predictions, one must consider the probabilistic nature of the models, and exercise care in basing decisions on predictions for a small number of compounds.

At the hit triage stage, where the task is an assessment of the likely properties of a series of compounds (either those already identified, or a virtual library of compounds that might be synthesized), assessment of the predictions over a larger number of compounds of more varied structure may give greater confidence than predicting for a single compound. Models which can be analyzed to identify correlations between physical properties or specific structural features can also be utilized as an input for hypothesis-generation for the synthesis of the next iteration of compounds.

5 Summary: Decision Making

When collected across a set of compounds from a chemical series, data on the properties described above can provide a composite picture of those compounds. As Fig. 10 illustrates, this data package can be daunting for any one compound; a potentially incomplete set of data on multiple compounds can be even more challenging to utilize. However, it is in building this network of information and then identifying key gaps in data and knowledge that teams effectively accomplish hit-triage. This composite picture should provide the basis for forming specific hypotheses for structural changes that can address the shortcomings of a particular series, and the comparison of property profiles and hypotheses across multiple chemical series should provide the basis for prioritization. The factors weighing

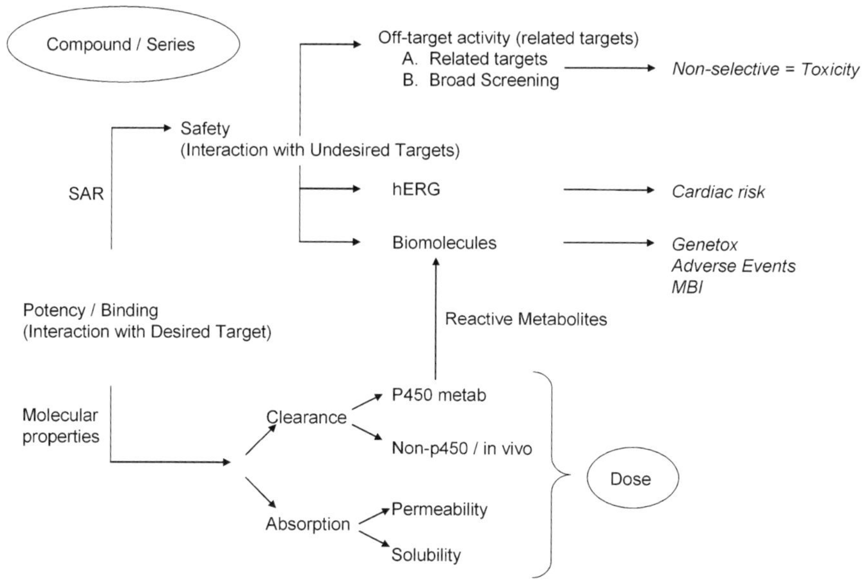

Fig. 10 A schematic of the information collected at hit-triage, and its relationship to relevant endpoints (dose and toxicity)

into the hit triage prioritization decision are varied but may include the expected ease with which the issues can be effectively addressed. Teams can also mitigate risk somewhat by balancing across a project so that chemical series with different issues are pursued in parallel. At the project level, the probabilities of success for each approach must be weighed such that a sufficient number of series are pursued such that the team has confidence that optimization of one will lead to the identification of a drug candidate.

References

1. Gillespie P, Goodnow RA Jr (2004) The hit-to-lead process in drug discovery. In: Doherty A (ed) Annual reports in medicinal chemistry. Elsevier, Oxford, p 293
2. Keseru GM, Makara GM (2006) Drug Discov Today 11:741
3. Leach AR, Hann MM, Burrows JN, Griffen EJ (2006) Mol Biosyst 2:429
4. Bleicher KH, Bohm H-J, Muller K, Alanine AI (2003) Nat Rev Drug Discov 2:369
5. Williams DH, Stephens E, O'Brien DP, Zhou M (2004) Angew Chem Int Ed Engl 43:6596
6. Nassar A-EF, Lopez-Anaya A (2004) Curr Opin Drug Discov 7:126
7. Lundqvist T (2005) Curr Opin Drug Discov 8:513
8. Carr RA, Congreve M, Murray CW, Rees DC (2005) Drug Discov Today 10:987
9. Stewart L, Clark R, Behnke C (2002) Drug Discov Today 7:187
10. Goodwill KE, Tennant MG, Stevens RC (2001) Drug Discov Today 6:S113
11. Rees DC, Congreve M, Murray CW, Carr R (2004) Nat Rev Drug Discov 3:660

12. Laird ER, Blake JF (2004) Curr Opin Drug Discov 7:354
13. Klebe G (2006) Drug Discov Today 11:580
14. Zartler ER, Mo H (2007) Curr Top Med Chem 7:1592
15. Hajduk PJ, Gerfin T, Boehlen J-M, Haeberli M, Marek D, Fesik SW (1999) J Med Chem 42:2315
16. Dalvit C, Flocco M, Knapp S, Mostardini M, Perego R, Stockman BJ, Veronesi M, Varasi M (2002) J Am Chem Soc 124:7702
17. Liu B, Li S, Hu J (2004) Am J Pharmacogenomics 4:263
18. Boozer C, Kim G, Cong S, Guan HW, Londergan T (2006) Curr Opin Biotech 17:400
19. Rich R, Hoth LR, Goeghegan KF, Brown TA, LeMotte PK, Simons SP, Hensley P, Myszka DG (2002) Proc Natl Acad Sci USA 99:8562
20. Cooper MA (2006) Drug Discov Today 11:1068
21. Doyle ML (1997) Curr Opin Biotech 8:31
22. Weber PC, Salemme FR (2003) Curr Opin Struct Biol 13:115
23. Dowling MR, Charlton SJ (2006) Br J Pharmacol 148:927
24. Huber W (2005) J Mol Recognit 18:273
25. Cik M, Jurzak MR (2006) High-throughput and high content screening. In: Taylor JB, Triggle DJ (eds) Comprehensive medicinal chemistry II. Elsevier, Oxford, Sect 3.32
26. Schnurr B, Ahrens T, Regenass U (2006) Optical assays in drug discovery In: Taylor JB, Triggle DJ (eds) Comprehensive medicinal chemistry II. Elsevier, Oxford, Sect 3.27
27. Ullmann D (2006) Fluorescence screening assays. In: Taylor JB, Triggle DJ (eds) Comprehensive medicinal chemistry II. Elsevier, Oxford, Sect 3.28
28. Weissman A, Keefer J, Miagkov A, Sathyamoorthy M, Perschke S, Wang FL (2006) Cell-based screening assays. In: Taylor JB, Triggle DJ (eds) Comprehensive medicinal chemistry II. Elsevier, Oxford, Sect 3.29
29. Kuntz ID, Sharp KA, Kollman PA (1999) Proc Natl Acad Sci USA 96:9997
30. Hopkins AL, Groom CR, Alex A (2004) Drug Discov Today 9:430
31. Hajduk PJ (2006) J Med Chem 49:6972
32. Lipinski CA, Lombardo F, Dominy BW, Freeny PJ (1997) Adv Drug Deliv Rev 23:3
33. Lipinski CA (2001) J Pharmacol Toxicol 44:235
34. Hajduk PJ, Greer J (2007) Nat Rev Drug Discov 6:211
35. Teague SJ, Davis AM, Leeson PD, Oprea T (1999) Angew Chem Int Ed Engl 38:3743
36. Hann MM, Leach AR, Harper G (2001) J Chem Inf Comput Sci 41:856
37. Proudfoot JR (2002) Bioorg Med Chem Lett 12:1647
38. Wunberg T, Hendrix M, Hillisch A, Lobell M, Meier H, Schmeck C, Wild H, Hinzen B (2006) Drug Discov Today 11:175
39. Hann MM, Oprea TI (2004) Curr Opin Chem Biol 8:255
40. Van de Waterbeemd H, Smith DA, Beaumont K, Walker DK (2001) J Med Chem 44:1313
41. Van de Waterbeemd H, Smith DA, Jones BC (2001) J Comp Aid Mol Des 15:273
42. Riley RJ, Martin IJ, Cooper AE (2002) Curr Drug Metab 3:527
43. Venkatakrishnan K, von Moltke LL, Obach RS, Greenblatt DJ (2003) Curr Drug Metab 4:423
44. Di L, Kerns EH (2005) Curr Opin Drug Discov Dev 8:495
45. Parmentier Y, Bossant M-J, Bertrand M, Walther B (2006) In vitro studies of drug metabolism. In: Taylor JB, Triggle DJ (eds) Comprehensive medicinal chemistry II. Elsevier, Oxford, Sect 5.10
46. Trager WF (2006) Principles of drug metabolism 1: redox reactions. In: Taylor JB, Triggle DJ (eds) Comprehensive medicinal chemistry II. Elsevier, Oxford, Sect 5.05
47. Testa B (2006) Principles of drug metabolism 2: hydrolysis and conjugation reactions. In: Taylor JB, Triggle DJ (eds) Comprehensive medicinal chemistry II. Elsevier, Oxford, Sect 5.06
48. Lin JH, Wong BK (2002) Curr Drug Metab 3:623
49. Obach RS (2001) Curr Opin Drug Discov Dev 4:36

50. Barter ZE, Bayliss MK, Beaune PH, Boobis AR, Carlile DJ, Edwards RJ, Houston JB, Lake BG, Lipscomb JC, Pelkonen OR, Tucker GT, Rostami-Hodjegan A (2007) Curr Drug Metab 8:33
51. Brown HS, Griffin M, Houston JB (2007) Drug Metab and Dispos 35:293
52. Grime K, Riley RJ (2006) Curr Drug Metab 7:251
53. Nassar A-EF, Kamel AM, Clarimont C (2004) Drug Discov Today 9:1020
54. Balimane PV, Han Y-H, Chong S (2006) AAPS J 8:E1
55. Artursson P, Neuhoff S, Matsson P, Tavelin S (2006) Passive permeability and active transport models for the prediction of oral absorption. In: Taylor JB, Triggle DJ (eds) Comprehensive medicinal chemistry II. Elsevier, Oxford, Sect 5.11
56. Kansy M, Senner F, Gubernator K (1998) J Med Chem 41:1007
57. Galinis-Luciani D, Nguyen L, Yazdanian M (2007) J Pharm Sci 96:2886
58. Avdeef A, Bendels S, Di L, Faller B, Kansy M, Sugano K, Yamauchi Y (2007) J Pharm Sci 96:2893
59. Yu X-Q, Xue CC, Wang G, Zhou S-F (2007) Curr Drug Metab 8:787
60. Fearn RA, Hirst BH (2006) Environ Toxicol Pharmacol 21:168
61. Abbott NJ (2006) In vitro models for examining and predicting brain uptake of drugs. In: Taylor JB, Triggle DJ (eds) Comprehensive medicinal chemistry II. Elsevier, Oxford, Sect 5.13
62. Malkia A, Murtomaki L, Urtti A, Kontturi K (2004) Eur J Pharm Sci 23:13
63. Alsenz J, Kansy M (2007) Adv Drug Deliv Rev 59:546
64. Bergstrom CAS, Wassvik CM, Johansson K, Hubatsch I (2007) J Med Chem 50:5858
65. Kola I, Landis J (2004) Nat Rev Drug Discov 3:711
66. Pearlstein R, Vaz R, Rampe D (2003) J Med Chem 46:2017
67. Raschi E, Vasina V, Poluzzi E, De Ponti F (2008) Pharmacol Res 57:181
68. Snyders DJ, Chaudhury A (1996) Mol Pharmacol 49:949
69. Kiehn J, Lacerda AE, Wible B, Brown AM (1996) Circulation 94:2572
70. Dubin AE, Nasser N, Rohrbacher J, Hermans AN, Marrannes R, Grantham C, Van Rossem K, Cik M, Chaplan SR, Gallacher D, Xu J, Guia A, Byrne NG, Mathes C (2005) J Biomol Screen 10:168
71. Raab CE, Butcher JW, Connolly TM, Karczewski J, Yu N, Staskiewicz S, Liverton N, Dean DC, Melillo DG (2006) Bioorg Med Chem Lett 16:1692
72. Singleton DH (2007) J Med Chem 50:2391
73. Fernandez D, Ghanta A, Kauffman GW, Sanguinetti MC (2004) J Biol Chem 279:10120
74. Cavalli A, Poluzzi E, De Ponti F, Recanatini M (2002) J Med Chem 45:3844
75. Aronov A (2006) J Med Chem 49:6917
76. Cianchetta G, Li Y, Kang J, Rampe D, Fravolini A, Cruciani G, Vaz RJ (2005) Bioorg Med Chem Lett 15:3637
77. Song M, Clark M (2006) J Chem Inf Model 46:392
78. Waring MJ, Johnstone C (2007) Bioorg Med Chem Lett 17:1759
79. Muto S, Miyata K, Tsutsumibata C, Daigo H, Ooro T, Sugiura K, Baba H, Uno Y (2006) Top Issues Appl Microbiol Biotech 1
80. Kirpnick Z, Homiski M, Rubitski E, Repnevskaya M, Howlett N, Aubrecht J, Schiestl RH (2005) Mutat Res 582:116
81. Avlasevich SL, Bryce SM, Cairns SE, Dertinger SD (2006) Environ Mol Mutagen 47:56
82. Greene N (2006) Computational models to predict toxicity. In: Taylor JB, Triggle DJ (eds) Comprehensive medicinal chemistry II. Elsevier, Oxford, Sect 5
83. Egan WJ, Zlokarnik G, Grootenhuis PDJ (2004) Drug Discov Today Tech 1:381
84. Contrera JF, Matthews EJ, Kruhlak NL, Benz RD (2005) Regul Toxicol Pharmacol 43:313
85. Murray M (2005) Mod Biopharm 4:1603
86. Hayashi M, Kamata E, Hirose A, Takahashi M, Morita T, Ema M (2005) Mutat Res 588:129
87. Matthews EJ, Kruhlak NL, Cimino MC, Benz RD, Contrera JF (2006) Regul Toxicol Pharmacol 44:97

88. Didziapetris R, Reynolds DP, Japertas P, Zmuidinavicius D, Petrauskas A (2006) Curr Comp Drug Des 2:95
89. Snyder RD (2007) Mutat Res 623:72
90. Hop CECA, Kalgutkar A, Soglia J (2006) Annu Rep Med Chem 41:369
91. Williams DP, Park BK (2003) Drug Discov Today 8:1044
92. Walgren JL, Mitchell MD, Thompson DC (2005) Crit Rev Toxicol 35:325
93. Park BK, Kitteringham NR, Maggs JL, Pirmohamed M, Williams DP (2005) Annu Rev Pharmacol Toxicol 45:177
94. Roth RA, Luyendyk JP, Maddox JF, Ganey PE (2003) J Pharmacol Exp Ther 307
95. Kalgutkar AS, Obach RS, Maurer TS (2007) Curr Drug Metab 8:407
96. Blagg J (2006) Annu Rep Med Chem 41:369
97. Soglia JR, Harriman SP, Zhao S, Barberia J, Cole MJ, Boyd JG, Contillo LG (2004) J Pharm Biomed Anal 36:105
98. Zheng J, Ma L, Xin B, Olah T, Griffith Humphreys W, Zhu M (2007) Chem Res Toxicol 20:757
99. Poulain R, Horvath D, Bonnet B, Eckhoff C, Chapelain B, Bodinier M-C, Deprez B (2001) J Med Chem 44:3391
100. Fliri AF, Loging WT, Thadeio PF, Volkmann RA (2005) Proc Natl Acad Sci USA 102:261
101. Schnecke V, Bostrom J (2006) Drug Discov Today 11:43
102. Beresford AP, Segall M, Tarbit MH (2004) Curr Opin Drug Discov Dev 7:36
103. Ekins S, Swaan PW (2004) Development of computational models for enzymes, transporters, channels, and receptors relevant to ADME/tox. In: Lipkowitz KB, Larter R, Cundari TR (eds) Reviews in computational chemistry, vol. 20. Wiley, Hoboken, NJ, chap 6
104. Penzotti J, Landrum GA, Putta S (2004) Curr Opin Drug Discov Dev 7:49
105. Bergstrom CAS (2005) Expert Opin Drug Metab Toxicol 1:613
106. Yu H, Adedoyin A (2003) Drug Discov Today 8:852
107. Davies JW, Glick M, Jenkins JL (2006) Curr Opin Chem Biol 10:343
108. Faller B, Ertl P (2007) Adv Drug Deliv Rev 59:533
109. Delaney JS (2005) Drug Discov Today 10:289
110. Bergstrom CAS (2005) Expert Opin Drug Metab Toxicol 1:613
111. Arimoto R (2006) Curr Top Med Chem 6:1609
112. Snyder RD, Smith MD (2005) Drug Discov Today 10:1119
113. Clark DE (2005) Computational prediction of blood–brain barrier permeation. In Doherty A (ed) Annual reports in medicinal chemistry. Elsevier, Oxford, p 413
114. Hutzler JM, Messing DM, Wienkers LC (2005) Curr Opin Drug Discov Dev 8:51
115. Ekins S, Balakin KV, Savchuk N, Ivanenkov Y (2006) J Med Chem 49:5059
116. Johnson SR, Yue H, Conder ML, Shi H, Doweyko AM, Lloyd J, Levesque P (2007) Bioorg Med Chem 15:6182

Top Med Chem (2009) 5: 175–212
DOI: 10.1007/7355_2009_2

Published online: 30 June 2009

Lead Identification

John W. Ellingboe and Adam M. Gilbert

Abstract High quality leads provide the foundation for the discovery of successful clinical development candidates, and therefore the identification of leads is an essential part of drug discovery. Many factors contribute to the quality of a lead, including biological, physicochemical, ADME, and PK parameters. The identification of high quality leads, which are needed for successful lead optimization, requires the optimization of all of these parameters. Parallel optimization of all parameters is the most efficient way to achieve the goal of lead identification.

Keywords Hit, Lead, Pharmacokinetics, Physicochemical

Contents

J.W. Ellingboe (✉) and A.M. Gilbert
Medicinal Chemistry, Wyeth Research, 401 N. Middletown Rd., Pearl River, NY 10965, USA
e-mail: ellingj@wyeth.com(Elligboe), gilbera@wyeth.com(Gilbert)

Abbreviations

Aβ	β-Amyloid
AcpS	Acyl carrier protein synthase
AD	Alzheimer's disease
ADME	Absorption distribution metabolism excretion
AGP	Alpha glycoprotein
APP	Amyloid precursor protein
BBB	Blood brain barrier
BBB PAMPA	Blood brain barrier parallel artificial membrane permeability assay
BSA	Bovine serum albumin
CCR1	Chemokine (C–C motif) receptor 1
CDK2	Cyclin-dependent kinase 2
CGRP	Calcitonin gene related peptide
CHK-1	Checkpoint kinase 1

CL_{int}	Intrinsic clearance
COST	Changing one separate factor at a time
$cPLA_2\alpha$	Cytosolic phospholipase $A_2\alpha$
CYP450	Cytochrome P450
DMSO	Dimethyl sulfoxide
DPP-4	Dipeptidyl peptidase IV
FACS	Fluorescence activated cell sorter
FLIPR	Fluorimetric imaging plate reader
GLP-1	Glucagon-like peptide-1
GPCR	G-protein coupled receptor
HCV	Hepatitis C virus
hERG	Human ether-a-go–go-related gene
HLM	Human liver microsomes
HMQC	Heteronuclear multiple quantum correlation
HSA	Human serum albumin
HTS	High throughput screen
IKKβ	IκB kinase-β
IP	Intellectual property
ITK	Interleukin-2-inducible T cell kinase
iv	Intravenous
LC/MS	Liquid chromatography/mass spectrometry
LE	Ligand efficiency
MCH1	Melanin-concentrating hormone 1
MIC	Minimum inhibitory concentration
MVA	Multivariate data analysis
MW	Molecular weight
NHR	Nuclear hormone receptor
NLM	Nonlinear mapping
PAMPA	Parallel artificial membrane permeability assay
PCA	Principal component analysis
PDE5	Phosphodiesterase 5
PGP	*P*-Glycoprotein
PK	Pharmacokinetic
PKCθ	Protein kinase C θ
PLS	Projection to latent structures by means of a partial least squares analysis
po	Oral
PSA	Polar surface area
RARβ2	Retinoic acid β2
RLM	Rat liver microsomes
SAR	Structure activity relationship
SPR	Structure property relationship
SPR	Surface plasmon resonance
STD	Saturation transfer difference

1 Introduction

High quality leads provide the foundation for the discovery of successful clinical development candidates, and therefore the identification of leads is an essential part of drug discovery. The process for the identification of leads generally starts with the screening of a compound collection, either an HTS of a relatively large compound collection (hundreds of thousands to one million plus compounds) or a more focused screen of a smaller set of compounds that have been preselected for the target of interest. Virtual screening methods such as structure-based or pharmacophore-based searches can complement or replace one of the above approaches. Once hits are identified from one or more of these screening methods, they need to be thoroughly characterized in order to confirm activity and identify areas in need of optimization. Finally, once fully characterized hits are identified, preliminary optimization through synthetic modification is carried out to generate leads. Parallel optimization of all properties, including biological, physicochemical, and ADME is the most efficient approach to the identification of leads. Hit characterization is described in the previous chapter. The focus of this chapter is on hit optimization and the identification of leads. After a general overview of these processes, examples taken from the literature since 2001 will be used to illustrate specific points. There are also a number of excellent reviews covering the lead identification process [1–6].

2 Lead Definition

A lead is variously defined in the pharmaceutical industry as a compound derived from a hit with some degree of in vitro optimization (potency in primary assay, activity in functional and/or cellular assay), optimization of physical properties (solubility, permeability), and optimization of in vitro ADME properties (microsomal stability, CYP inhibition). Moreover, a lead must have established SAR/SPR around these parameters such that continued optimization appears possible. A lead may also have preliminary PK and in vivo animal model data. However, it is the task of the lead optimization chemist to improve PK and in vivo activity to the levels needed for identification of a clinical candidate.

3 Establish Lead Profile

At the outset of a lead identification effort, it is imperative to establish specific criteria for potency, selectivity, ADME properties, etc. to generate a desired lead profile. This profile serves to guide the lead identification efforts based on the initial characterization of the hits.

Criteria for biological properties may be project specific, but ADME property and physical property criteria are generally invariant. Lead profiles will be addressed in more detail in the section on parallel optimization.

4 Characterization of Hits

Hits need to be thoroughly characterized to remove compounds that do not interact with the molecular target in a specific manner, to prioritize the remaining compounds for follow up work, and to provide direction for optimization to leads. Characterization includes biological profiling, physicochemical and ADME profiling [7–9], and mechanistic profiling. The profile of the starting compounds ultimately determines the quality of the leads.

4.1 Biophysical Characterization/Enzymology

While it is important to confirm the activity of hits in the primary assay and also to collect cellular and/or functional, and ADME data, it is also essential to establish that the hits interact with the molecular target in a specific and stoichiometric fashion as functional optimization and establishment of real SAR is impossible with compounds that show nonspecific binding. It has been shown that one source of false positives in the screening of enzyme targets is inhibition by colloidal aggregates, which nonspecifically inhibit the enzyme [10]. Many techniques are available for demonstrating the specificity of binding of a compound to a molecular target [11–15].

4.2 Clustering, Series Formation or Identification of Singletons

Once all the data are collected, including potency, selectivity, functional, and ADME data, and evidence of binding to the molecular target, it is useful to start trying to define clusters or series of compounds by chemical structure, so that efforts can be focused in the areas holding most promise. Compounds can be clustered simply by visual inspection of structures or by using various clustering algorithms. There are algorithms based on ring scaffold analysis [16], or based on chemical fingerprints [17, 18]. Clustering may be purely based on chemical structure, or biological and other data may be introduced to start analyzing SAR. The goal of clustering the compounds is to identify compound series that demonstrate a robust SAR, in which compounds possess a range of potencies that can be explained by the variations in structural features. A cluster of hits may not show a distinct

SAR, but this may be because there are insufficient compounds or the compounds do not possess sufficient structural variation.

Efforts to cluster a set of compounds will usually also result in a number of singletons. While a lack of compounds with related structures having biological activity may seem discouraging, if the biological data for a singleton are sufficiently compelling, it is worth retaining such compounds and trying to build a series through synthesis.

4.3 Pharmacophore/Binding Model

As all of the data are collected and analyzed it is also important to start considering a binding model or pharmacophore model to help explain the variation in biological activity with structure, and to provide a basis for the design of new analogs.

For targets that lack structural information, such as GPCRs or ion channels, a pharmacophore model or multiple pharmacophore models for different series of compounds can explain SAR and guide the synthesis of new analogs. Alternatively, homology models based on bacteriorhodopsin have been used to explain the interactions of small molecules with GPCRs.

When an X-ray crystal structure for the protein molecular target has been solved, it is possible to model the interactions of the small molecule ligand with the target. Better yet are co-crystal structures of the ligand with the molecular target. However, because of the flexibility of proteins and the changes in three dimensional protein structure that can be induced by the binding of a ligand [19], it is ideal to have co-crystal structures for representative compounds from each series being followed up for lead identification.

4.4 Patentability Assessment

If the goal in a lead identification effort is to provide a patentable series of compounds, either at the outset of lead optimization or later during optimization, then an initial novelty assessment is appropriate once hit series are formed. Initial literature searches using a tool such as SciFinder can give a good indication of how crowded a structural class is in terms of prior art. Further refined searches of patent databases such as Marpat provide information on patents or patent applications that might generically cover a hit series. These searches may be delayed until some synthetic work has been completed and data generated so that a narrower, more focused search can be done. While patentability is an important consideration for lead identification efforts in pharmaceutical companies, it must be balanced with the biological and pharmaceutical data. It is possible that significant structural changes will be made during the lead optimization stage that will introduce novelty into a chemical series.

5 Supplementing Characterized Hits

In the process of clustering compounds to try to form series, and examining all of the data for the compound in a cluster, it may become apparent that there is an insufficient number of compounds, or enough range in the data to reach a decision as to whether a cluster of compounds is a series with demonstrated SAR. This problem can be addressed through searching of databases for additional related compounds or limited synthesis to provide the critical compounds.

5.1 Substructure and Similarity Searching

Substructure and similarity searching of compound collections not screened in the HTS campaign, including additional compounds in a corporate collection or commercially available collections, can provide additional compounds for testing. These compounds may help strengthen a cluster by generating additional data points and a more robust SAR. Similarity searching based on methods such as topological atom pairs [20] are related to the methods used for clustering.

5.2 Preliminary Array Synthesis

The synthesis of small, focused arrays based on some of the hits may help define a series if the desired compounds are not available through substructure or similarity searching. The purpose of this type of array is not to optimize a hit but to help prioritize a series for further synthetic work. Inclusion of the original hit in the array is useful for confirmation of activity, particularly when supplies are limited or the screening sample was of low purity.

6 Parallel Optimization

Just as it is important to characterize hits fully and to collect not only potency data but also physicochemical and ADME profiling data and evidence of specific binding to the molecular target, optimization of the hits to yield leads requires a focus on all of these parameters. It is most efficient – although also more challenging – to optimize biological, physicochemical, and ADME properties in parallel. If only biological potency is optimized, then other properties such as MW or lipophilicity may increase in an unwanted direction. If one then tries to correct MW or lipophilicity, then potency may be lost. High throughput assays are now available for many of the physicochemical and ADME properties detailed below, so most or all of the compounds synthesized in a lead identification campaign can be fully characterized.

As mentioned in Sect. 3, it is important to establish a detailed lead profile at the beginning of a lead identification effort. Criteria vary in different lead identification or hit-to-lead groups, but generally include some or all of the following: potency, functional activity, selectivity, MW, clogP, solubility, permeability, microsomal stability and/or hepatocyte clearance, and preliminary PK including oral bioavailability. An example of a lead profile for a kinase inhibitor project is illustrated in Table 1 [21].

Included in this table are criteria related to kinase inhibition, including detailed analyses of reversibility, detergent effects, and competition with ATP. Also listed are criteria for selectivity, cellular activity, and the physicochemical and in vitro ADME profiles. The final two criteria require the lead to be part of a series of compounds with demonstrated SAR.

A detailed screening pathway starting with a screen to generate hits and culminating in leads is also important for guiding lead generation efforts. An example is illustrated in Fig. 1.

Table 1 Example of desired profile for lead

	Desired profile
IC_{50} (μM)	<1.0 μM
Reversibility	>70%
10 × enzyme (IC_{50} change)	<fivefold
w/wo Triton (IC_{50} change)	<fivefold
Enzyme kinetics (K_i μM)	<10.0 μM
IC_{50}(FL) (μM) w/Triton	<1.0 μM
Binding to target (NMR, FP, Trp-Fl.)	Yes (NMR, FP)
Cell assay 1 (50% of control)	<10 μM
Cell assay 2	<10 μM
Selectivity enzyme 1	>10-fold
Selectivity other enzymes	As a bonus
MW	<450
clogP	<4.0
PSA	<80
Aqueous solubility (μg mL^{-1} at pH 7.4)	>60
Permeability (10^{-6} cm s^{-1} at pH 7.4)	>1
CYP 3A4 inhib (at 3 μM)	<15
CYP 2D6 inhib (at 3 μM)	<15
CYP 2C9 inhib (at 3 μM)	<15
Microsome stability (% remaining at 15 min)	>80
Preliminary PK Profile	Yes
hERG block (at 10 μM)	<20%
Definable series	Yes
Definable SAR	Yes

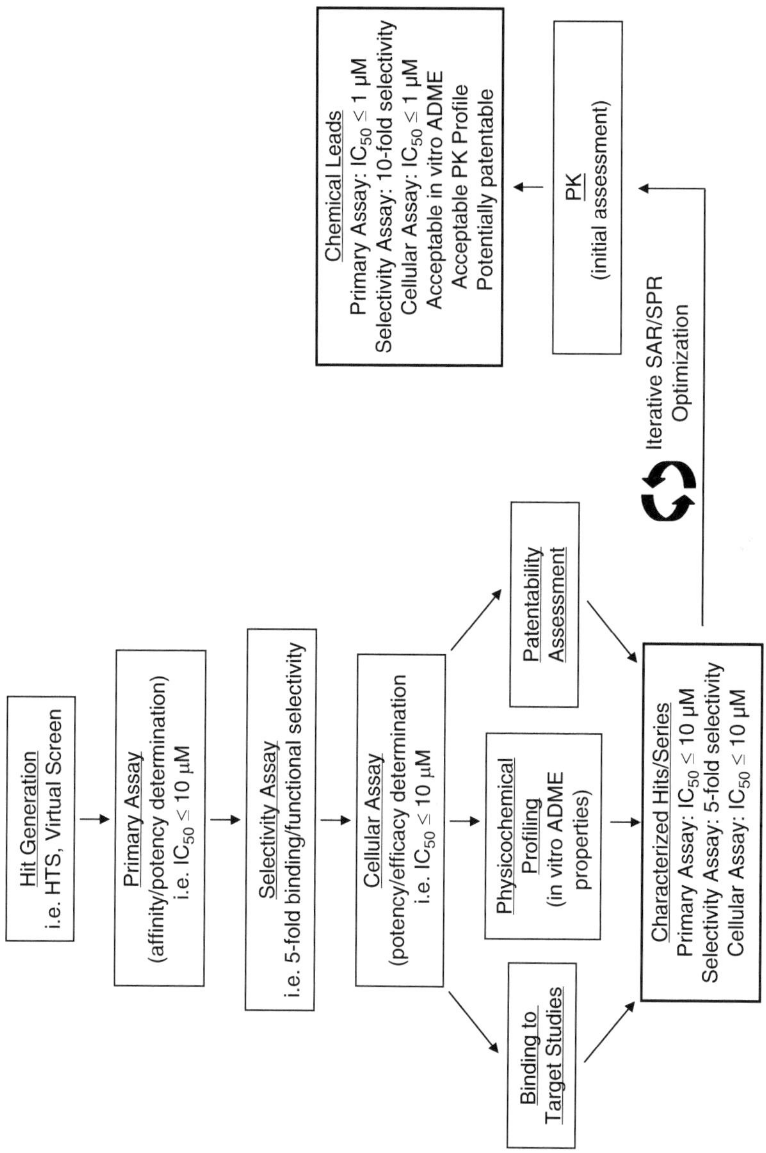

Fig. 1 Example of hit to lead screening pathway

6.1 Biological Properties

Generally, all of the assays used for characterization of hits in terms of potency, selectivity, function, and or cellular activity will continue to be used to characterize new compounds synthesized to identify leads. However, additional assays will normally be added to characterize compound selectivity more fully, to provide additional evidence that compounds are acting on the desired pathway and by the desired mechanism.

6.1.1 Potency

The primary assay used to generate hits will continue to be useful for characterizing the potency of new compounds against the molecular target. In some cases, however, the format used by an HTS group may be changed for a benchtop assay where a greater range of detection methods may be accommodated. Also, in some cases, while an HTS campaign may have used one type of assay, such as binding, to identify hits, during the lead identification phase a different assay type, such as a cellular assay, may become the primary assay. During the lead identification stage, achieving an increase in potency of at least one order of magnitude builds confidence in a series and suggests that it can be fully optimized during the lead optimization stage. For example, if a hit series has a potency range of 1–5 μM, then the synthesis of compounds with a potency in the 100–500 nM range would meet the potency criterion.

6.1.2 Selectivity

Starting with hits that may not be selective vs a nearest neighbor or vs another target associated with some liability, at the lead identification stage it is important to start focusing on introducing selectivity. The selectivity criterion will be project dependent, but a minimum of 10-fold selectivity is a good target for leads assuming this selectivity can be rationalized and there is a potential for further improving selectivity. It is expected that selectivity will be fully achieved during the lead optimization stage.

Additional selectivity assays for a wider range of molecular targets will also be introduced. These could include, for example, a panel of kinases representing the various classes of kinases or a panel of GPCRs.

6.1.3 Function

For programs where the primary assay is a binding type assay, it is important to continue to evaluate the functional response of newly synthesized compounds. Changes in functional potency should correlate with changes in binding potency. Other parameters that may need to be tracked are efficacy and agonism vs antagonism.

6.1.4 Cellular Activity

An assay for cellular activity of newly synthesized compounds is important because it can be used to demonstrate functional activity in a more complex system that is one step closer to an in vivo model. Again, changes in cellular activity should correlate with changes in potency in a binding or inhibition assay, although other physicochemical parameters such as solubility and permeability also need to be considered. Cellular assays can also confirm the mechanism of action of a compound on a pathway. For example, a phosphoblotting assay can be used to show that inhibition of a kinase blocks the phosphorylation of the kinase substrate specifically. Demonstration of activity in a cellular assay can be used as one criterion for selecting compounds to test in vivo models in the lead optimization stage, so it is important to have compounds fully characterized in this type of assay during the lead identification stage.

6.2 Physical Properties

During the characterization process, hits are typically tested for kinetic solubility and permeability in a model of passive diffusion such as PAMPA [22]. As new compounds are synthesized, additional parameters also need to be considered, such as p*K*a, chemical and plasma stability, and protein binding. Calculated properties such as MW, clogP, and PSA should also be tracked.

6.2.1 Solubility

All newly synthesized compounds should be tested for solubility to ensure that a series is maintaining good solubility properties, or in the case of hits with poor solubility, that there is an improvement in solubility. Good compound solubility is essential for data from all other assays to be meaningful, and it is important to maintain or introduce good solubility from the beginning of the drug discovery effort to ensure ultimately that clinical candidates have the good solubility that is important for developability [23]. In addition to routine kinetic solubility assays in a standard buffer, other custom assays may be needed. For example, if biological data do not follow a trend that can be rationalized, or there is a lack of correlation between binding and cellular assay results, it is worth checking the solubility of compounds in the various assay media used for a program. In some cases, compounds may exhibit quite different solubilities in different media.

When compounds are selected for preliminary PK studies, the identification of an appropriate dosing vehicle for iv studies requires solubility studies in various vehicles. Also, the study of thermodynamic solubility is useful as this more closely reflects the environment experienced by compounds on oral dosing. Higher throughput thermodynamic solubility assays have been introduced recently [23] so it will be possible to introduce this type of assay earlier in the discovery process.

6.2.2 p*K*a

The ionizability of compounds affects other parameters such as solubility, permeability, and ultimately oral bioavailability, so it may be important to track changes in the p*K*a of new compounds. Calculated p*K*a values can be used when planning the synthesis of new compounds, but it is also a good idea to confirm these values experimentally. An example where this strategy can be useful is in the search for bioisosteric replacements for a carboxylic acid group.

6.2.3 Stability (Chemical, Plasma)

The confirmation of the chemical and plasma stability of new compounds is important to ensure that the synthesized compounds are actually responsible for the observed biological activity (rather than a degradation product), and to help ensure good bioavailability. Selected compounds can be tested for chemical stability in buffers across a pH range of 1.0–9.0, and in simulated gastric and intestinal fluids. Also, assaying compounds for stability in plasma gives one an initial indication of how a compound may behave in an in vivo system. High throughput assays for plasma stability are available [24].

6.2.4 Protein Binding

The characterization of the protein binding of lead compounds is important because of the effect that high levels of protein binding can have on PK and in vivo model data, and ultimately on clinical efficacy. Only unbound drug is available to interact with the molecular target. However, a highly bound compound with a fast off rate may still exhibit good in vivo profiles. Binding to HSA and AGP is measured with equilibrium dialysis assays, or other higher throughput assays such as affinity chromatography with immobilized HSA [25]. A preliminary indication of protein binding can also be identified by running an in vitro biological assay in the absence and presence of protein such as BSA.

6.2.5 Calculated Properties

In addition to the in vitro assays described above, physical properties should be calculated for all new compounds designed for synthesis. It is necessary to keep in mind the target values for leads, such as MW < 450, clogP < 4.0, and PSA < 80. It has been demonstrated that properties such as MW and clogP increase during optimization [26], so that a lead needs to have lower values for these properties than a drug candidate. Additional factors that make up the rule-of-five [27] as well as the number of rotatable bonds as described by Veber [28] can also be tracked. While these properties should be kept under the upper limits because of the correlation with good PK properties [29], they may also be useful as variables when conducting multivariate data analyses of SAR. After compounds are synthesized and tested, LE, defined as the binding energy of the ligand per atom [30, 31], can also be tracked for each series.

6.3 *In Vitro ADME*

Drug candidates that are intended for oral dosing need to have good ADME properties so that they can be dosed once or twice daily. The drug should be well absorbed, survive first pass metabolism, and have sufficiently low clearance. At the lead identification stage, the primary in vitro ADME assays employed are those that assess permeability and metabolic stability. There are a variety of assays available for both parameters, as described in the previous chapter.

6.3.1 Metabolite Identification

If low metabolic stability is identified in a series, then it is useful to carry out metabolite identification in order to pinpoint the site of metabolism and to develop a synthetic strategy to address the stability. A general strategy for identifying the structure of metabolites is to incubate the compound of interest with liver microsomes to generate the metabolite and carry out a preliminary analysis using LC/MS techniques [32]. The LC/MS results may be sufficient to determine an exact structure, but in some cases additional analysis using NMR or LC/NMR may be required to establish an exact structure. If necessary, synthesis can be used to confirm a structural assignment, and to provide material for testing to determine if it is an active metabolite.

With the structure of a metabolite in hand, synthetic strategies to block the metabolism can be designed. Various synthetic modifications have been shown to block oxidation, dealkylation, glucuronidation, etc. [33]. Solving metabolic stability issues can be one of the tougher challenges early in the drug discovery process, but it is important to try to solve them early to help ensure that high quality drug candidates are delivered.

6.4 *Toxicology*

While little in vivo testing may take place in the lead identification stage, there are some in vitro approaches for looking at specific aspects of toxicology, such as inhibition of CYP450 enzymes which could lead to drug–drug interactions in humans, as described in the previous chapter, and hERG inhibition, which could lead to cardiac arrhythmias. At the lead identification stage, it is best to look at trends within series for these liabilities, and to try to eliminate common structural features within series or overall properties of series to correct the issues. Other types of assays such as incubation of compounds with hepatocytes have been used to look at to toxicity towards a specific organ type.

6.4.1 hERG Inhibition

Inhibition of potassium current through the I_{Kr} channel encoded by hERG can cause long QT syndrome, the prolongation of the QT signal in an electrocardiogram, and this can in turn result in potentially fatal torsade de pointes cardiac arrhythmias. A number of marketed drugs were withdrawn when it was found that they caused sudden cardiac death, and it is now essential that all drug candidates be free of hERG inhibition at the concentration likely to be found on dosing. The standard assay for assessing hERG inhibition is a patch-clamp. It is important at the lead generation stage to look for trends in series, and to try to eliminate structural features that my be responsible for hERG inhibition in a series [34]. There are a number of in silico approaches that can be used to try to predict hERG liabilities [35, 36], but in vitro assays are needed for confirmation.

7 Pharmacokinetics

During the lead identification process, PK data are used to varying degrees, depending on the availability of PK assay support for early drug discovery. Ideally, for a given series, it would be best to have PK data for several iterations of compounds, with the goal of improving PK properties if the initial hit has poor properties. A complete data set including both iv and po dosing is optimal when the target profile is oral dosing.

It is also important to correlate PK data with in vitro ADME assay data for each series to validate the predictive potential for the in vitro assays within a series. Once this is done, one can rely on the in vitro assays with more confidence.

With more limited PK support, an initial PK profile for a lead compound may be all that is available. This profile identifies issues that need to be solved during lead optimization.

There are a number of solutions that have been proposed to address the limitations on throughput in PK assays, including cassette dosing [37], where typically five compounds are dosed in a mixture, pooling of plasma samples from multiple animals receiving a specific dose, or the cassette-accelerated rapid rat screen where the processing of samples is streamlined [38].

In addition to the data generated from plasma samples in a standard PK study, for targets that are contained within the brain, it is also important to determine brain levels of compound, so that brain/plasma ratios can be determined. Again, these data should be correlated with in vitro assays and calculated properties such as BBB PAMPA [39] and PSA for a given series. For example, while the default cutoff PSA value for good BBB penetration is 80, for a specific series of compounds it may be lower.

8 Tools for Data Analysis

As discussed above, it is important to try to optimize biological, physicochemical, and ADME properties in parallel. However, the data from all of these assays for the numerous compounds prepared by parallel synthesis make the interpretation of results challenging. The use of tools such as MVA helps in the effective utilization of all data in the optimization process.

8.1 Multivariate Data Analysis

MVA is a very useful tool for classifying sets of compounds and identifying the primary latent variables that summarize the data through PCA, and for identifying correlations between variables describing the properties of compounds and the biological effects of these compounds through PLS [40].

PCA is a projection method that allows one to take a multivariate data matrix and represent it in low-dimension space. It then becomes more straightforward to identify dominant patterns and major trends in the data. The relationships between compounds and data, and among the data variables, are uncovered. In PCA we take linear combinations of observations (compounds) and variables. The data matrix is summarized row-wise as scores (t_a) and column-wise as loadings (p_a). The directionality in a scores plot corresponds to that of a loadings plot, so the dominant variables associated with a compound can be identified [41]. Through these techniques (PCA and PLS), correlations between molecular descriptors, measured physical properties, ADME assay results, and biological assay results can be uncovered in large data sets, which could not readily be found through visual inspection of data tables. This type of analysis can provide guidance for future iterations of synthesis.

8.2 Non-Linear Mapping

In addition to looking for data trends in physical property space using PCA and PLS, trends in chemical structure space can be delineated by viewing nonlinear maps (NLM) of two-dimensional structure descriptors such as Unity Fingerprints or topological atom pairs using tools such as Benchware™ DataMiner [42]. Two-dimensional NLM plots provide an overview of chemical structure space and biological activity/molecular properties are mapped in a 3rd and/or 4th dimension to look for trends in the dataset.

9 Tools for the Design of Synthetic Targets

Multiple approaches are available to aid in the design of new analogs to optimize hits with the goal of identifying leads. Depending on the target type and the availability of structural information, either structure-based or pharmacophore-based design of new analogs may be appropriate. In the initial phases of lead identification, the synthesis of several arrays of compounds based on each hit can help to determine quickly whether the hit is a singleton or whether a series can be formed around the hit. These arrays can also provide some initial SAR and guidance on how to focus future iterations.

9.1 Structure-Based, Structure-Guided Array Synthesis

For soluble protein targets such as enzymes or protein–protein interactions the availability of structural information can greatly aid the design of new analogs. In the lead identification stage, while potency improvement is one goal that can be addressed with structural information, the larger challenges are often selectivity, physical properties, and ADME issues. With a co-crystal structure of a hit bound to the molecular target, it is possible to identify regions of the hit that could accommodate, for example, a functional group that would improve solubility without interfering with the interactions that contribute to compound binding affinity for the target. Similarly, areas where metabolism-blocking groups can be introduced without affecting binding can be identified. Finally, with structures of selectivity targets in hand, it is possible to start introducing structural changes that may improve selectivity.

It is particularly effective to pair structure-based design with array synthesis. A relatively larger virtual library of possible analogs can be docked into the structure and the analogs prioritized for synthesis or eliminated if scored low.

If crystal structures of the molecular target and co-crystal structures of bound ligands are not available, the use of homology models may help with analog design. Homology models based on related enzymes may help with design, but may also need to be refined further based on assay data of new analogs. Homology models based on bacteriorhodopsin have been used to rationalize the SAR of compounds synthesized for GPCR targets.

9.2 Pharmacophore Guided Array Synthesis

The design of new analogs based on pharmacophore models is complementary to structure-based design. This method may be used even where structural information is available, but is also of use for targets where structural information is not available, such as GPCRs. A pharmacophore model can provide guidance about which functional groups in a structure are essential for binding, and which areas may be open for greater variation to optimize physical properties, etc.

9.3 Design of Experiments Applied to Array Design

When optimizing a hit with more than one region that can be varied, it is important not to vary only one group at a time while holding the others constant. The strategy of COST does not necessarily lead to the optimal compound. Systems influenced by more than one factor are poorly studied by the COST strategy [43]. To avoid missing an optimal compound without having to synthesize every possible combination of substituents (which could require the synthesis of many thousands of compounds), the use of a Design of Experiments (DoE) approach is both efficient and can lead one towards the optimal combination of substituents. Using DoE, one avoids synthesizing redundant compounds and obtains the maximum amount of SAR information through the synthesis of a relatively small number of representative compounds. Also, a designed set of compounds can result in a better PLS model. While DoE was initially used primarily for optimization in process research, more recently it has been demonstrated to have a variety of applications in drug discovery [44], including the design of arrays [45].

9.4 Scaffold Hopping

For various reasons, including the potential for patentability, chemical tractability, or structural features that contribute to physicochemical or ADME issues, it may be desirable to identify a new scaffold for lead identification in a program. Retaining appendages that may be key for interaction with the molecular target, the core or scaffold in a hit can be replaced with a new scaffold that yields compounds with improved properties [46]. New scaffolds can be identified computationally through similarity searches of compound collections, or through de novo design, using structural information if available.

10 Establishing an Intellectual Property Position

Finally, in addition to addressing all the scientific issues outlined above during the lead identification process, the design of analogs with features that ensure the potential patentability of a series is an important consideration for lead identification work done in a commercial enterprise. The timing of the filing of patent applications will depend on the overall patenting strategy of the company where the work is carried out. If patents are filed at the lead stage, then a detailed patentability assessment is required. If patent filing takes place later in the lead optimization stage, it is still important to consider the potential for patenting a series, including analogs that are anticipated to be made in the future, when recommending a lead for optimization.

11 Illustrations with Examples

The following examples taken from the literature since 2001 are intended to illustrate the parallel optimization of some or all of the parameters discussed above, and to illustrate how some of the tools described above can aid in lead generation. It is not meant to be an exhaustive survey of the literature. The specific criteria used to define a successful lead identification campaign vary by group, as do the processes used to reach the lead stage. However, there are also many themes that are common to many of the examples below. A summary of the examples described below is contained in Table 2.

11.1 Example 1: CCR4 Antagonists

CCR4 is a member of the GPCR CC chemokine family predominantly expressed on T cells, and is involved in the migration of inflammatory cells. Antagonists have the potential to treat inflammatory diseases. Wang et al. [47] report on the optimization of a μM naphthalene/thiazole hit which has a high iv clearance and a short in vivo half-life. After optimizing the linker between the naphthalene and the thiazole (NH is optimal), optimization of the 4-thiazole position to bulky alkyl groups provided compounds with increased CCR4 potency using a TARC (thymus and activation regulated chemokine) binding assay. Naphthalene replacements were identified (*N*-methyl indole) as well as thiazole replacements (*N*-methyl pyrazole and pyridine cores). However the optimal compound maintains the original naphthalene amine-substituent, the original thiazole core and incorporation of an *N, N*-bis(cyclohexylmethyl)methyl amine moiety at the 4-thiazole position (Scheme 1). This optimized compound shows improved potency over the starting hit, improvement of cellular activity in a CEM cell migration assay, and improved clearance and half life in a PK study.

11.2 Example 2: cPLA$_2\alpha$ Inhibitors

cPLA$_2\alpha$ releases arachidonic acid to initiate the production of multiple mediators of inflammation, such as leukotrienes, prostaglandins, and thromboxanes. Inhibition of this enzyme has the potential to be useful for treating inflammatory diseases. McKew et al. [48] report on the optimization of a weakly potent benzhydryl indole diacid with weak rat whole blood activity measuring the inhibition of thromboxane B$_2$ production. Incorporation of a 2-indole methyl group to increase potency,

Table 2 Summary of lead identification examples

Entry	Author	Target	Features optimized	Tools used	Reference
1	Wang	CCR4	Potency, cellular, PK		[47]
2	McKew	$cPLA_2\alpha$	Potency, cellular, PK		[48]
3	Souers	MCH1	Potency, cellular, PK	B/P ratio	[49]
4	Lund	RARβ2	Potency, cellular, physicochem, PK		[50]
5	Beaulieu	HCV pol	Potency, physicochem	NMR binding	[51]
6	Williams	CGRP	Potency, cellular, PK	Pharmacophore-based design	[52]
7	Morwick	IKKβ	Potency, physicochem, ADME	Pharmacophore model, scaffold hopping	[53]
8	Baxter	IKK-β	Potency, cellular, physicochem, ADME, PK	Homology model	[54]
9	Cywin	PKCθ	Potency, cellular, physicochem	Homology model	[55]
10	Poulain	μ Opioid	Potency, cellular, physicochem	Pharmacophore-based design	[56]
11	Gilbert	AcpS	Potency, cellular, physicochem, ADME	Crystallography, PCA, PLS, DoE	[45]
12	Palmer	PDE5	Potency, physicochem, ADME	LE, crystallography	[57]
13	Baxter	CXCR2	Potency, cellular, physicochem, ADME, PK		[58]
14	Baxter	CXCR2	Potency, cellular, physicochem, ADME, PK		[59]
15	Ho	CXCR2	Potency, ADME		[60]
16	Armour	CCR5	Potency, selectivity, physicochem, ADME, PK	Modeling of CYP 2D6	[61]
17	Pevarello	CDK2	Potency, selectivity, physicochem, ADME, PK	Crystallography	[62]
18	Baxter	$P2X_7$	Potency, physicochem, ADME, PK		[63]
19	Backes	DPP-4	Potency, physicochem, ADME, PK	Crystallography, scaffold hopping	[64]
20	Cole	BACE-1	Potency, cellular	Crystallography	[65]
21	Wang	mGluR1	Physicochem, ADME, PK		[66]
22	Snow	ITK	Potency, selectivity		[67]
23	Xie	CCR1	Potency		[68]
24	Tong	CHK-1	Potency	Crystallography	[69]

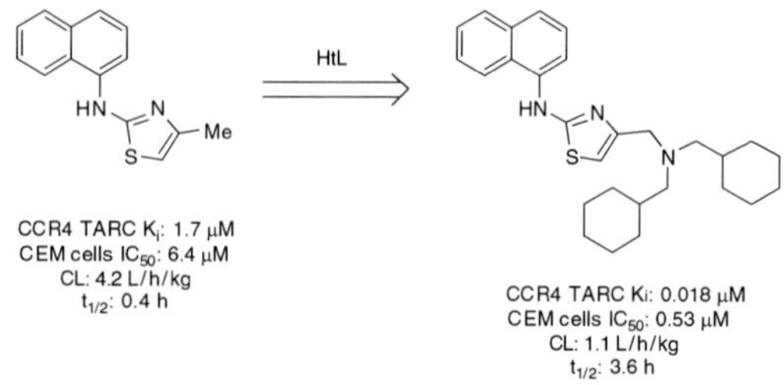

Scheme 1 CCR4 antagonists

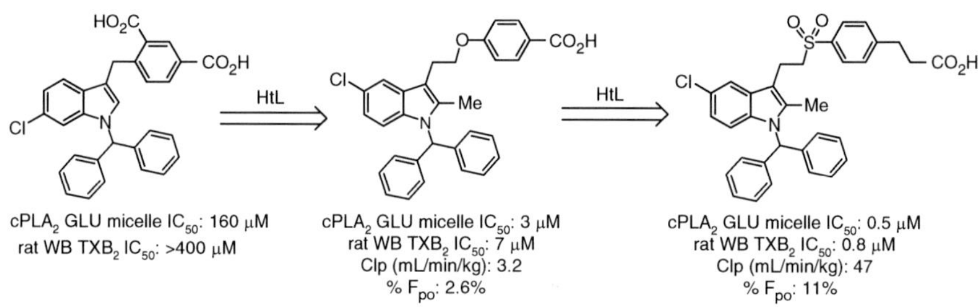

Scheme 2 cPLA$_2\alpha$ inhibitors

removal of one of the carboxylic acids as well as elongation of the linker between the 3-indole and benzoic acid moiety gives compounds that show improved potency in a cPLA$_2$ GLU micelle assay as well as μM-potency in a rat whole blood assay. Additional optimization of the linker and extension of the acid moiety away from the benzene ring gives a sub-μM compound even in the whole blood assay. Although the optimized compound shows relatively high clearance, oral bioavailability has been improved (Scheme 2).

11.3 *Example 3: MCH1 Receptor Antagonists*

MCH is a cyclic 19-amino acid peptide that plays a major role in body weight regulation in rodents. Antagonism of the MCH1 receptor is a potential approach to the treatment of obesity. Souers et al. [49] report on the optimization of a 2-aminoquinoline HTS hit which initially showed divergent MCH1R binding affinity and MCH1R FLIPR potency. Optimization of the oxygen CyHCH$_2$-substituent

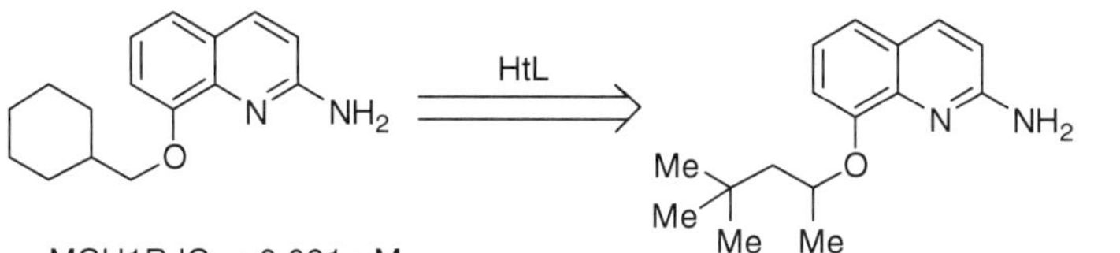

Scheme 3 MCH1 receptor antagonists

shows that increased MCH1R binding affinity and potency can be obtained by using hydrophobic/bulky moieties (i.e. a 4,4-dimethylpentan-2-ol group). Variation of the 2-amino group produces compounds with reduced affinity and MCH1R antagonism. The optimized lead was found to have potent and correlated MCH1R affinity/functional antagonism, showing good exposure after po dosing at 10 mg kg^{-1} and an excellent B/P ratio at 10 mg kg^{-1}, po (Scheme 3).

11.4 Example 4: RARβ2 Receptor Agonists

The retinoic acid β2 receptor is a nuclear receptor, and agonism has the potential for treatment of cancer and other hyperproliferative disorders. Lund et al. [50] report on the optimization of AC-55649, a sub-μM agonist of RARβ2 with high cLogD and poor aqueous solubility. RARβ2 agonism was increased by incorporating a F atom *ortho-* to the CO_2H moiety. The physicochemical profile of the lead was improved by converting one of the phenyl rings to a thiazole and by transforming the octyl-alkyl chain to a 2-butoxyethyl moiety. The resulting lead, AC-261066, displays an RARβ2 $pEC_{50} = 8.0$, an aqueous solubility of 4.8 mg mL^{-1}, a clogD = 0.7 and an % F = 52%. A homology model based on RARγ was utilized in the optimization (Scheme 4).

11.5 Example 5: HCV NS5B Polymerase Inhibitors

HCV NS5B polymerase is an RNA-dependent RNA polymerase that is essential for viral replication. Thus, the inhibition of this enzyme offers a potential treatment for hepatitis C infection. Beaulieu et al. [51] report on the parallel optimization of enzyme inhibition potency and physical properties. In the first stage of hit characteri-

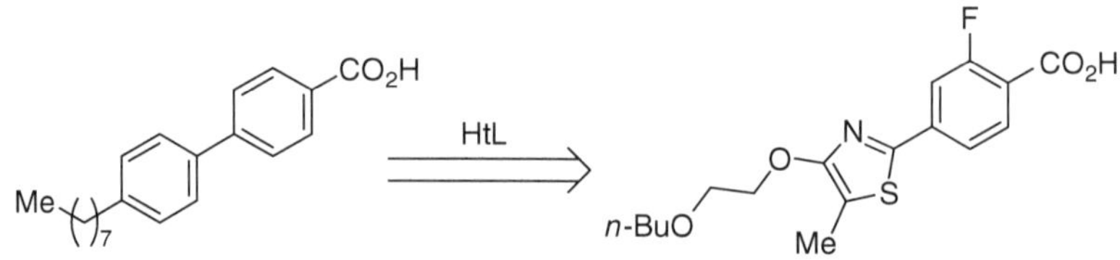

AC-55649
RARβ2 pEC$_{50}$: 6.9 (92% eff)
sol < 0.001 mg/mL
cLogD 5.6

AC-261066
RARβ2 pEC$_{50}$: 8.0 (96% eff)
sol 4.8 mg/mL
cLogD 0.7
%F$_{po}$: 52%

Scheme 4 RARβ2 receptor agonists

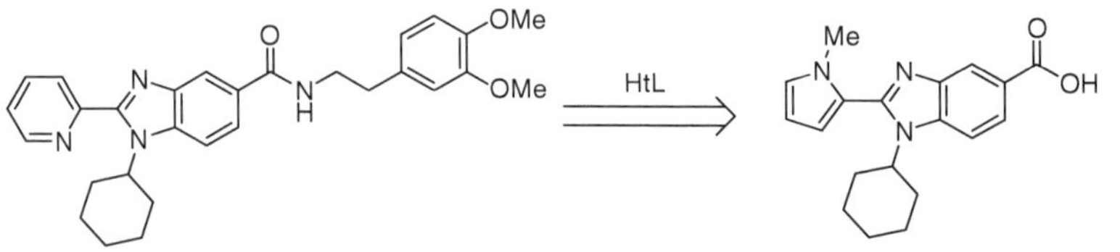

HCV NS5B IC$_{50}$: 12 μM
Poliovirus IC$_{50}$: > 500 μM
MW: 484.59

HCV NS5B IC$_{50}$: 1.8 μM
Poliovirus IC$_{50}$: > 500 μM
MW: 323.39

Scheme 5 HCV NS5B polymerase inhibitors

zation, the group employed NMR differential line broadening and transferred-NOESY experiments to confirm the specificity of ligand binding to the target protein. In the course of optimization, the right side of the hit was truncated resulting in a MW reduction from 484 to 323 by converting the 3,4-dimethoxyphenethyl amide to a carboxylic acid, which also reduces the clogP and TPSA of the lead. Unfortunately, the compounds did not show activity in a replicon assay (Scheme 5).

11.6 *Example 6: CGRP Antagonists*

CGRP is a 37 amino acid neuropeptide that is a potent vasodilator, and CGRP release is associated with migraine headaches. Antagonism of the CGRP receptor is an attractive target for the treatment of migraines. Williams et al. [52] describe the optimization of a μM CGRP antagonist lead by modifying the spiro[imidazolidinedione]tetrahydronaphthalene to a piperidinyl-benzoimidazolone which mimics the same moiety in Boehringer–Ingelheim CGRP antagonist BIBN 4096 BS. The benzodiazepine-1-methyl substituent was also transformed to a 2,2,2-trifluoroethyl moi-

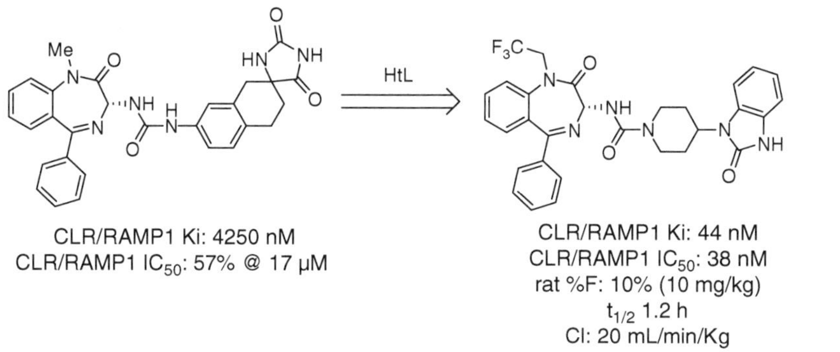

Scheme 6 CGRP antagonists

ety. These changes increase potency by 100-fold, lower the TPSA, and produce a compound with good PK parameters. The CGRP receptor is heterodimeric, and the assay utilized a cell line with both components (CLR and RAMP1) expressed or a human neuroblastoma cell line expressing native receptor. A pharmacophore model was used to guide synthesis, and the authors postulate that the piperidinyl–benzoimidazolone group acts as a constrained Ala–Phe–NH_2 moiety (Scheme 6).

11.7 *Example 7: IKKβ Inhibitors*

IκB kinase-β is a key regulatory enzyme in the NF-κB pathway, and inhibition of this enzyme has the potential for yielding treatments for inflammatory and autoimmune diseases. Morwick et al. [53] report on the optimization of a μM IKKβ inhibitor with low aqueous solubility, moderate human liver microsome stability, and inhibition of several CYPs (3A4, 2C9, 1A2) with μM potencies. Modulation of the thiophene core (other thiophene isomer, pyrimidine and oxazole) produces compounds of similar potency to the hit. Fusing the 5-phenyl moiety to the thiophene to form a thieno[2,3-b]pyridine core increases aqueous solubility of the series as well as reduces the CYP liability. While the optimized compound still shows μM IKKβ potency, the aqueous solubility, HLM stability and CYP profiles are much improved. A pharmacophore model was generated that enabled scaffold hopping to yield this new chemotype (Scheme 7).

11.8 *Example 8: IKKβ Inhibitors*

In a second example of the identification of IKKβ inhibitor leads (termed IKK2 in this paper), Baxter et al. [54] report on the optimization of enzyme and cellular potency, physicochemical properties, ADME properties, and PK. This group targets

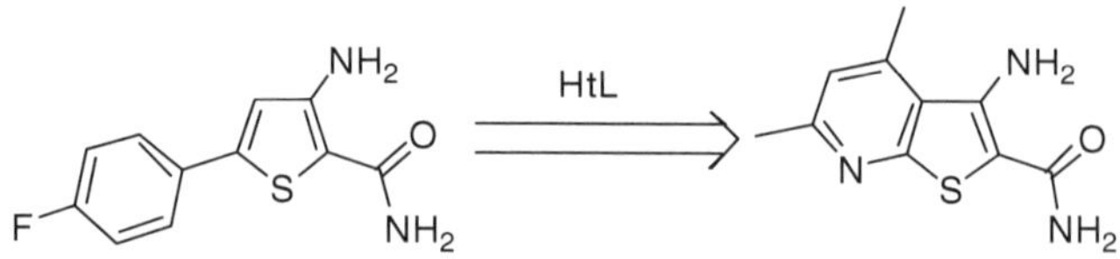

IKKβ IC_{50}: 2.6 μM
IKKα: 24 μM
Aq. sol. pH 7.4: 7 μg/mL
Caco-2 A to B: 19.3 cm/s
HLM $t_{1/2}$: 12 min
CYP IC_{50}s: 2 to >30 μM

IKKβ IC_{50}: 2.2 μM
IKKα: 2.8 μM
Aq. sol. pH 7.4: 24 μg/mL
Caco-2 A to B: 13.6 cm/s
HLM $t_{1/2}$: 78 min
CYP IC_{50}s: >30 μM

Scheme 7 IKKβ inhibitors (1)

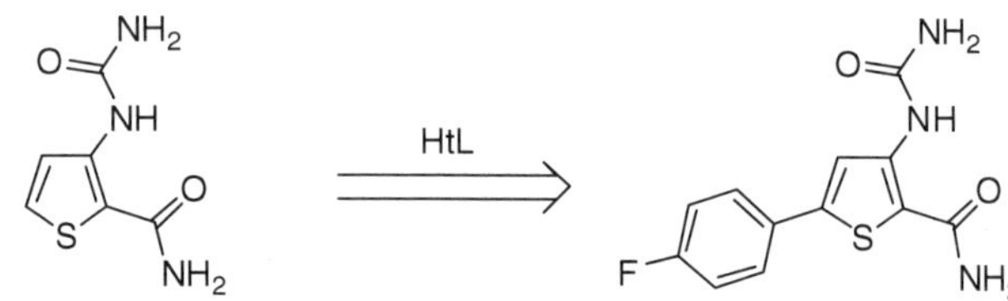

IKK-2 IC_{50}: 2 μM
IKK-1 IC_{50}: N/A
TNF cell IC_{50}: 10 μM
Rat Hepatocyte Cl: 14 μl/min/10^6 cells
HLM Cl: 49 μl/min/mg

IKK-2 IC_{50}: 0.06 μM
IKK-1 IC_{50}: 6.3 μM
TNF cell IC_{50}: 0.40 μM
Rat Hepatocyte Cl: 2 μl/min/10^6 cells
HLM Cl: 8 μl/min/mg
Rat Cl_{iv}: 6 ml/min/kg
Rat % F_{po}: 78
Plasma protein binding: 92.5%

Scheme 8 IKKβ inhibitors (2)

rat hepatocyte CL_{int} of <14 μL min^{-1} 10^6 $cells^{-1}$ and human liver microsome clearance of <23 μL min^{-1} mg^{-1} (which represent half of the hepatic blood flow) for all of their Hit to Lead activities. This example is similar to 11.7 above except that the thiophene core does not have a 4- or 5-substituent. The main issue with the μM IKKβ inhibitor in Scheme 8 appears to be the high rat hepatocyte and HLM clearance which is likely due oxidation of the thiophene scaffold. Appending a (4-F)Ph group to the 5-position of the thiophene not only significantly improves the IKKβ potency, but also improves the CL_{int}. The optimized chemical lead also has excellent rat oral bioavailability and is not highly bound to plasma protein.

11.9 Example 9: PKCθ Inhibitors

Protein kinase C theta plays a critical role in T cell signaling and thus the inhibition of this enzyme has the potential to be useful for treating inflammatory diseases. Cywin et al. [55] describe the parallel optimization of potency against the enzyme,

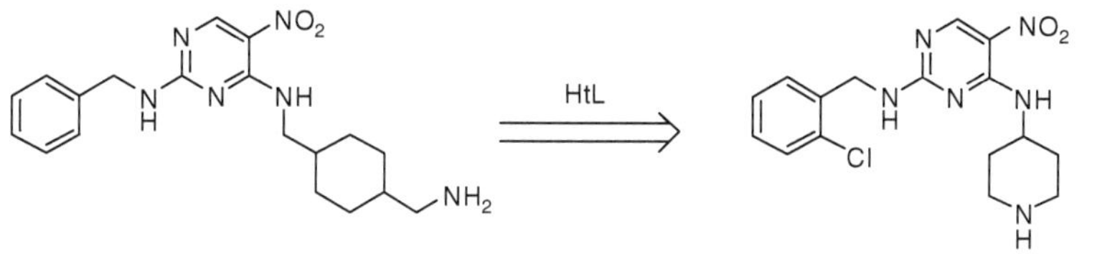

PKCΘ IC_{50}: 0.5 μM

PKCΘ IC_{50}: 0.09 μM
IL-2 IC_{50}: 0.72 μM
CYP2C9 IC_{50}: 17.6 μM
CYP2D6 IC_{50}: >30 μM
CYP3A4 IC_{50}: 28 μM
HLM $t_{1/2}$: 66 min
Caco-2 AB: 8.1 x 10^{-6} cm/sm

Scheme 9 PKCθ inhibitors

potency in a cellular assay for IL-2 production, physicochemical, and ADME properties. A homology model based on CDK2 was used to rationalize SAR which led to the exploitation of the 2-benzylamino substituent in order to build in PKCθ potency as well as selectivity vs VEGFR1, LYN, IR, and SYK kinases. Also of importance is the transformation of the (4-cyclohexyl)methanamine moiety to a 4-piperidine which maintains PKCθ potency, improves the CYP profile and most likely improves the in vivo clearance of the series (no glucuronidation of the primary amine). It is noted in the paper that the lead contains a nitro group, which would raise a toxicity structural alert (Scheme 9).

11.10 Example 10: μ Opioid Receptor Modulators

The μ opioid receptor is a GPCR and a potential target for pain. In a study designed to illustrate two approaches to pharmacophore-based design for optimization, Poulain et al. [70] describe the optimization of binding affinity and physicochemical properties. Potency at the μ, κ, and δ opioid receptors is increased by basifying the central amine of the starting hit by removing the carbamate functionality. Replacement of the nitro group with a less problematic chlorine atom improves both opioid receptor potency as well as aqueous solubility. Key in the optimization is the replacement of the piperidinylbenzimidazolone fragment with a phenyltriazaspirodecanone moiety which shows similar μ and κ potency and improved selectivity over δ. Aqueous solubility can also be improved by the incorporation of the *N*-methylacetamide group (Scheme 10).

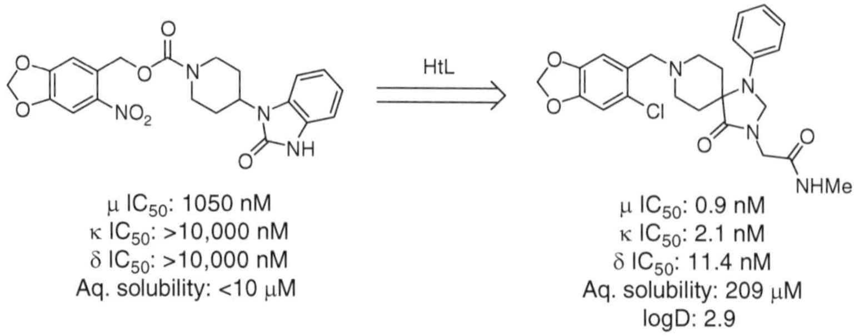

Scheme 10 μ Opioid receptor modulators

11.11 Example 11: AcpS Inhibitors

AcpS is an essential enzyme in bacteria that is required for the biosynthesis of components of membrane lipids and bacterial lipopolysaccharides, and is thus an attractive antibacterial target. Gilbert et al. [45] describe the parallel optimization of enzyme inhibition potency, antibacterial efficacy (minimum inhibitory concentration or MIC), and physicochemical properties. In this work, an initial array of 42 4*H*-oxazol-5-ones was prepared using Design of Experiments (DoE) software where three *R*-groups were varied around a 2-phenyl-4-((phenylamino)methylene)oxazol-5(4*H*)-one scaffold. MVA (PLS) of the calculated physical properties of these *R*-groups, the in vitro AcpS IC_{50} data as well as the MIC antibacterial potency shows a clear trend where the 2-(4-(trifluoromethyl)phenyl)oxazol-5(4*H*)-one moiety gives increasing AcpS potency and a slight improvement in antibacterial activity. A second array was next prepared again using DoE software varying additional *R*-groups where the 2-(4-(trifluoromethyl)phenyl)oxazol-5(4*H*)-one moiety was incorporated into every product. Optimized compounds were identified (an example is presented in Scheme 11) where AcpS has been improved to sub-μM potency and antibacterial activity had been improved as well.

11.12 Example 12: PDE5 Inhibitors

PDE5 cleaves cGMP, and inhibition of this enzyme helps to maintain cGMP concentrations, which in turn relaxes blood vessels. Inhibitors of this enzyme are marketed for erectile dysfunction. Palmer et al. [57] describe the parallel optimization of enzyme inhibition potency, selectivity, physicochemical properties and PK. The initial start point was a potent PDE5 inhibitor with an unacceptably high MW for a hit to lead starting point. A key PDE5 fragment with reduced MW (309 vs 545 for the hit) and LE (0.4 vs 0.27 for the hit) was identified via preparation of a library of 192 compounds aided by a PDE5/sidenafil crystal structure. Modification of the

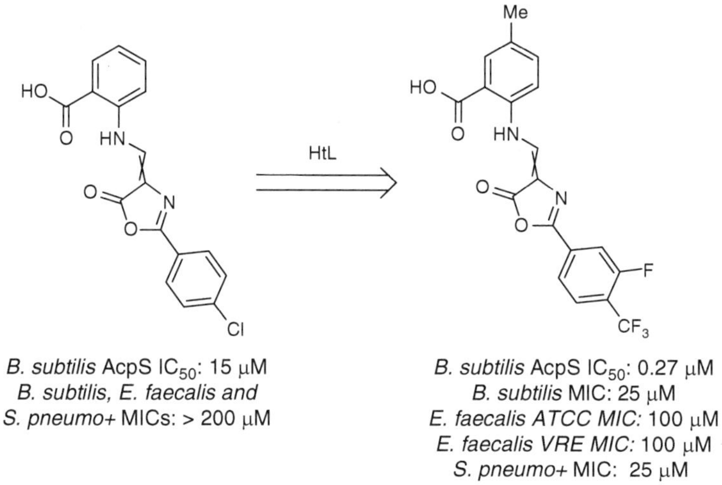

Scheme 11 AcpS inhibitors

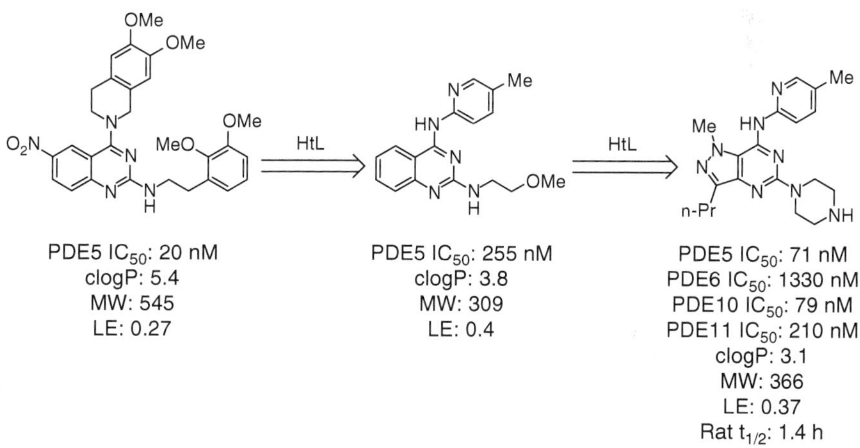

Scheme 12 PDE5 inhibitors

quinazoline scaffold to a 1-methyl-3-propyl-1*H*-pyrazolo[4,3-*d*]pyrimidine and constraining the 2-methoxyethylamine to a piperidine produces an optimized lead with potent PDE5 potency, good selectivity over PDE6, a clogP and MW that would be consistent with a brain permeable compound and a good in vivo half-life in rat (Scheme 12).

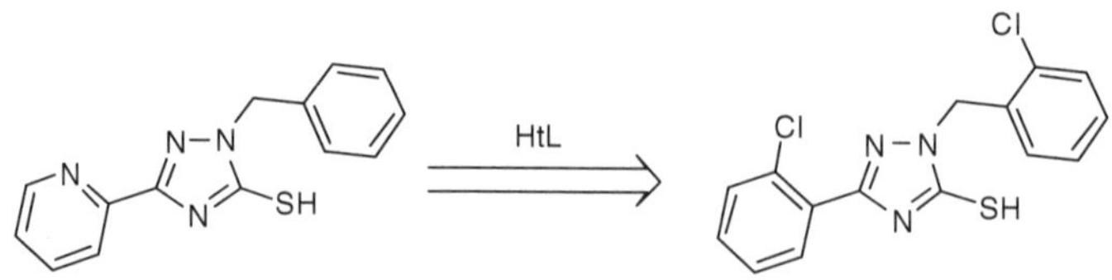

CXCR2 Binding IC_{50}: 4.6 μM
CXCR2 FLIPR IC_{50}: 2.4 μM
MW: 268
clogP: 3.4
Rat Hepatocyte Cl: 19μl/min/10^6 cells
HLM Cl: 13 μl/min/mg
Rat Cl_{iv}: 12 ml/min/kg

CXCR2 Binding IC_{50}: 0.028 μM
CXCR2 FLIPR IC_{50}: 0.048 μM
MW: 336
clogP: 4.3
sol: 20 μg/ml
Rat Hepatocyte Cl: 26 μl/min/10^6 cells
HLM Cl: 14 μl/min/mg
Rat Cl_{iv}: 12 ml/min/kg
Rat $\%F_{po}$: 61%
Plama proten binding: 99%

Scheme 13 CXCR2 antagonists (1)

11.13 Example 13: CXCR2 Antagonists

CXCR2 is a member of the CXC family of chemokine receptors. IL-8 activates this receptor, and an antagonist would potentially be useful for the treatment of inflammatory diseases. Baxter et al. [58] describe the parallel optimization of binding and functional potency, physicochemical properties, ADME properties, and PK. The thiol of the HTS hit was varied with typical replacements (i.e., OH, NH_2, SMe, NHAc, etc.), but this only led to inactive compounds. Variation of the substituent at N(2) showed that a benzyl moiety was required (Ph, Me substituents gave inactive compounds). Variation of the C(5) substituent showed that *o*-substituents produced optimal activity. The optimized lead has substantially improved CXCR2 binding and functional activity as well as an excellent PK profile (Scheme 13).

11.14 Example 14: CXCR2 Antagonists

In a second example of the identification of a CXCR2 antagonist lead, Baxter et al. [59] discuss the parallel optimization of binding potency and functional potency, physicochemical properties, ADME properties, and PK for a different chemotype. This initial hit showed high rat hepatocyte clearance presumably due to the aromatic OH and NH_2 groups as well as the lipophilic *n*-pentyl group. The 2-, 5- and 7-positions of the thiazolo[4,5-*d*]pyrimidine scaffold were sequentially varied to optimize potency. The 2-amino group could be changed to a hydrogen, however this group was maintained as an NH_2 presumably since it confers good aqueous

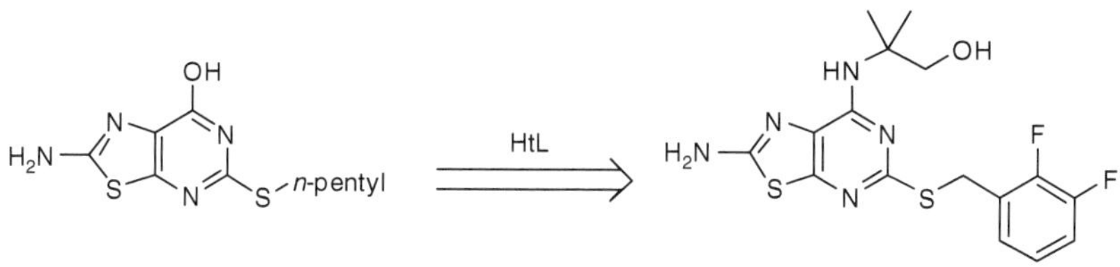

CXCR2 Ki: 10 μM	CXCR2 Ki: 0.014 μM
CXCR2 IC_{50}: 2 μM	CXCR2 IC_{50}: 0.04 μM
Rat hepatocyte Cl: 49 μl/min/10^6 cells	Rat hepatocyte Cl: 4 μl/min/10^6 cells
HLM Cl: 18 μl/min/10^6 cell	HLM Cl: 31 μl/min/10^6 cell
MW: 270	MW: 397
Aq. solubility: 27 μg/mL	Aq. solubility: 0.5 μg/mL
clogP: 2.0	clogP: 3.1
logD: 2.9	logD: 3.4
	Rat Cl_{iv}: 25 mL/min/kg
	Rat_{iv} $T_{1/2}$: 1.2 h
	Rat $\%F_{po}$: 15%
	Plasma protein binding: 98.4%

Scheme 14 CXCR2 antagonists (2)

solubility to the series. The 5-thio *n*-pentyl group could be modified to a benzyl moiety. The incorporation of fluorines on the benzyl group adds additional potency to this series. Finally the 7-hydroxyl group could be modified to a much more hindered 2-amino-2-methylpropan-1-ol. The resulting lead showed increased potency and substantially improved rat hepatocyte clearance. For this lead, most criteria were met, with the exception of solubility, which decreased (Scheme 14).

11.15 Example 15: CXCR2 Antagonists

A third example of the identification of a CXCR2 antagonist lead is described by Ho et al. [60]. This group carried out the parallel optimization of binding affinity, ADME properties, and PK. While the initial hit had good CXCR2 potency, stability in RLM was only modest due to the highly lipophilic S-*n*-octyl substituent on the pyrimidine. Shortening of the S-*n*-octyl group led to improvement of rat liver microsomal stability but at the expense of CXCR2 potency. CXCR2 potency could be gained back by the addition of a phenyl moiety at the 4-position of the imidazole group. The optimal balance of potency and stability was obtained with a pyrimidine S-*n*-propyl substituent in combination with an imidazole *p*-CF_3OPh moiety (Scheme 15).

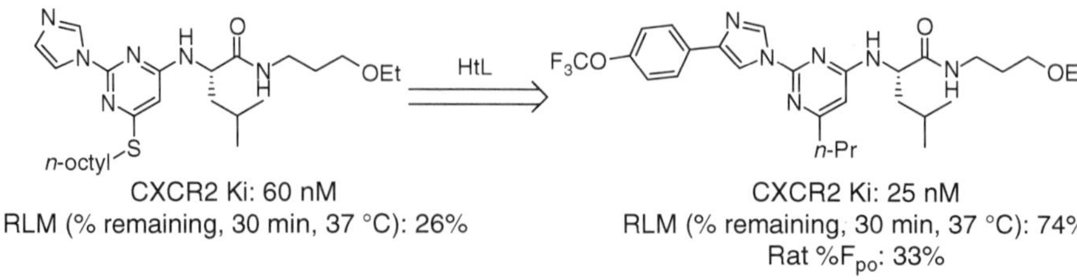

Scheme 15 CXCR2 antagonists (3)

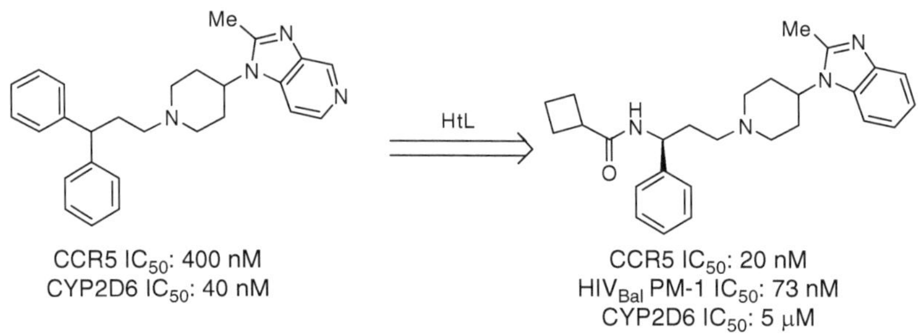

Scheme 16 CCR5 antagonists

11.16 Example 16: CCR5 Antagonists

The CCR5 chemokine receptor is a GPCR and a member of the CC family. It is a major coreceptor for fusion and entry of HIV into cells, and is thus an attractive target for the treatment of HIV infection. Armour et al. [61] describe the parallel optimization of binding potency, efficacy (inhibition of HIV replication in PM-1 cells), physicochemical and ADME properties. The initial hit was a strong inhibitor of the CYP450, CYP2D6, and modeling of the CYP2D6 enzyme was used to identify a way to reduce this inhibition (Scheme 16).

11.17 Example 17: CDK2 Inhibitors

CDK2 is involved with controlling normal cell proliferation. Disregulation in cancer makes this a good antitumor target. Pevarello et al. [62] describe the parallel optimization of enzyme inhibition potency, cellular activity, physicochemical properties, and PK. A low MW hit (MW = 201) was specifically selected with the

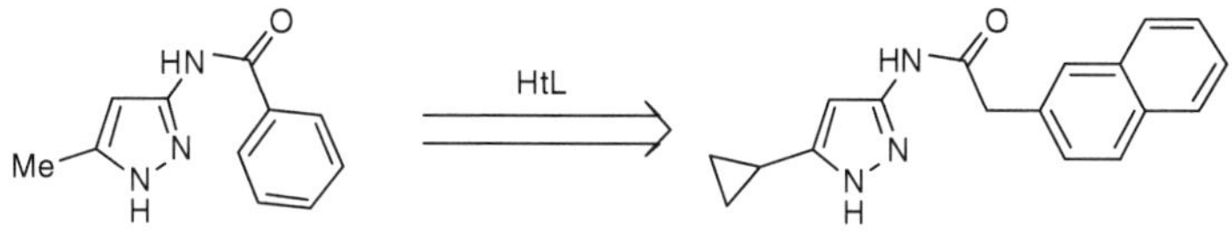

MW 201
CDK2/cyclinA IC_{50}: 1500 nM

MW 292
CDK2/cyclinA IC_{50}: 37 nM
A2780 cells IC_{50}: 286 nM
HT-29 cells IC_{50}: 170 nM
HTC116 cells IC_{50}: 73 nM
DU145 cells IC_{50}: 651 nM
>50 fold selective vs. 33 kinases except:
CDK2/cyclinE (2.5 fold), CDK5/p25 (3.1 fold),
CDK1/cyclinB (7.3 fold)
Rat Cl_{iv}: 13.5 mg/min/kg (5 mg/kg)
Rat $\%F_{po}$: 90% (10 mg/kg)

Scheme 17 CDK2 inhibitors

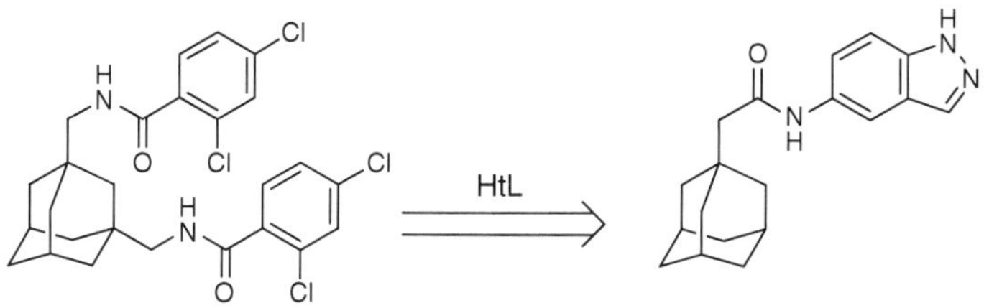

$P2X_7$ pA_2: 6.9
MW: 540
clogP: 6.9

$P2X_7$ pA_2: 7.4
MW: 309
clogP: 4.1
Rat Hepatocyte Cl: 5 μl/min/10^6 cells
HLM Cl: 21 μl/min/mg
Rat Cl_{iv}: 47 ml/min/kg

Scheme 18 $P2X_7$ inhibitors

goal of identifying a low MW lead, in this case the lead MW is 292. Crystallography was used to guide the design of new analogs (Scheme 17).

11.18 Example 18: $P2X_7$ Inhibitors

The $P2X_7$ receptor is a ligand-gated ion channel present in cells involved with inflammation. The receptor is activated by extracellular ATP, which leads to the processing and release of IL-1β. Baxter et al. [63] report on the parallel optimization of binding affinity, efficacy, physicochemical properties, ADME properties, and PK (Scheme 18).

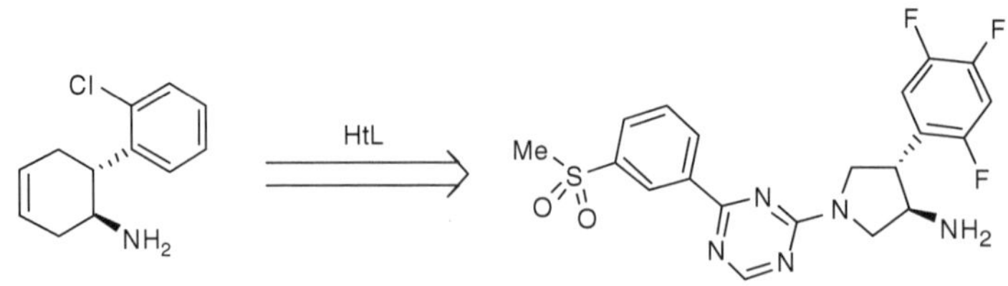

DPP4 Ki: 0.8 μM

DPP4 Ki: 0.002 μM
>1000-fold selective vs. DPP8, DPP9, POP and FAPα
IC_{50} > 10 μM vs. CYP3A4, CYP2D6, CYP2C9
RLM and HLM $t_{1/2}$: > 3h
Rat Cl_{iv}: 0.15 L/h/kg (5 mg/kg)
Rat $\%F_{po}$: 74% (5 mg/kg)
Plasma protein binding: < 95%

Scheme 19 DPP-4 inhibitors

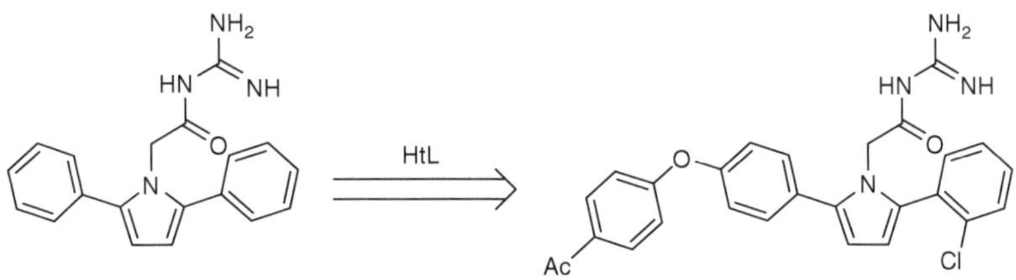

BACE-1 IC_{50}: 3.7 μM

BACE-1 IC_{50}: 0.7 μM
BACE-2 IC_{50}: 2.3 μM
CathD IC_{50}: 18.9 μM
Cell Aβ total ED_{50}: 3.8 μM

Scheme 20 BACE-1 inhibitors

11.19 Example 19: DPP-4 Inhibitors

DPP-4 is a serine protease that inactivates GLP-1. GLP-1 stimulates insulin secretion and suppresses glucagon release. The inhibition of DPP-4 prolongs the half-life of GLP-1 and brings about beneficial effects on glucose levels and glucose tolerance in type 2 diabetics. Backes et al. [64] report on the parallel optimization of enzyme binding affinity and inhibition, selectivity, ADME properties, and PK (Scheme 19).

11.20 Example 20: BACE-1 Inhibitors

BACE-1 (β-secretase) is one of the enzymes involved in breaking down APP to produce Aβ (amyloid β-peptide, $A\beta_{40,42}$), the protein that eventually oligomerizes to form Aβ plaques, the hallmark of Alzheimer's disease (AD). Thus an agent that

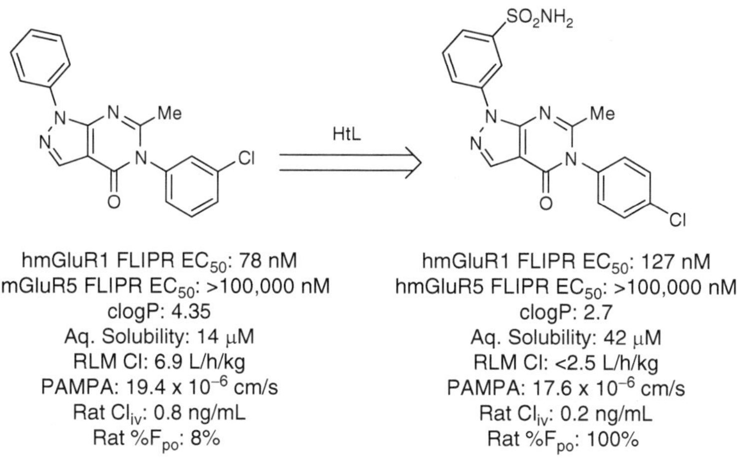

Scheme 21 mGluR1 antagonists

inhibits BACE-1 could be a disease-modifying treatment for AD. Cole et al. [65] report on the optimization of a μM acylguanidine BACE-1 inhibitor to a sub-μM compound with some selectivity over BACE-2, good selectivity over Cathepsin D and low μM activity in a cellular assay of Aβ production. X-ray crystallography of the initial hit was used to find substituents for the two pyrazole phenyl rings to explore effectively the S_1 and $S_{2'}$ pockets (Scheme 20).

11.21 Example 21: mGluR1 Inhibitors

The Group I metabotropic glutamate receptors located in the dorsal horn have been well validated as targets for pain. In particular, mGluR1 antagonists have become an attractive target for nociception. Wang et al. [66] have shown that slight changes to an HTS hit can lead to large improvements in the PK profile of a series. Adjustment of the phenyl groups at the 1 and 5 positions of a 1*H*-pyrazolo[3,4-d] pyrimidin-4(5*H*)-one scaffold does not change the excellent mGluR5/mGluR1 functional selectivity of the series, but greatly improves the rat microsomal CL_{int} (6.9 to < 2.5 L h^{-1} kg^{-1}), and also improves in vivo rat clearance (0.8 to 0.2 ng mL^{-1}) and oral bioavailability (8 to 100%) (Scheme 21).

11.22 Example 22: ITK Inhibitors

Interleukin-2-inducible T cell kinase (ITK) is expressed mainly in T-lymphocytes and plays a major role in the activation of T-cells. Thus inhibitors of ITK should be useful as immunosuppressives and antiinflammatory agents. Snow et al. [67] describe a novel series of 2-aminobenzimidazole ITK inhibitors which is optimized

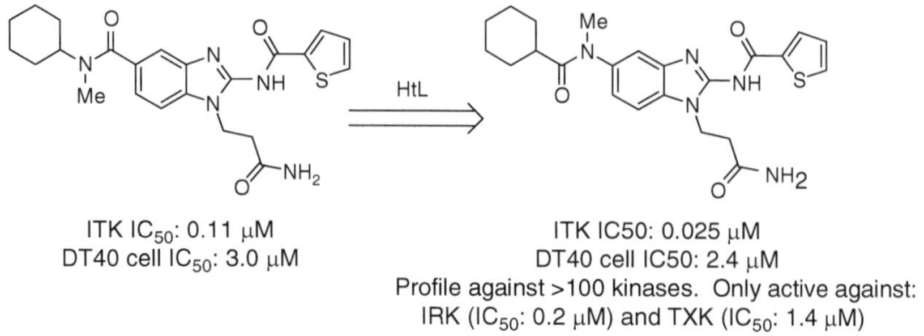

Scheme 22 ITK inhibitors

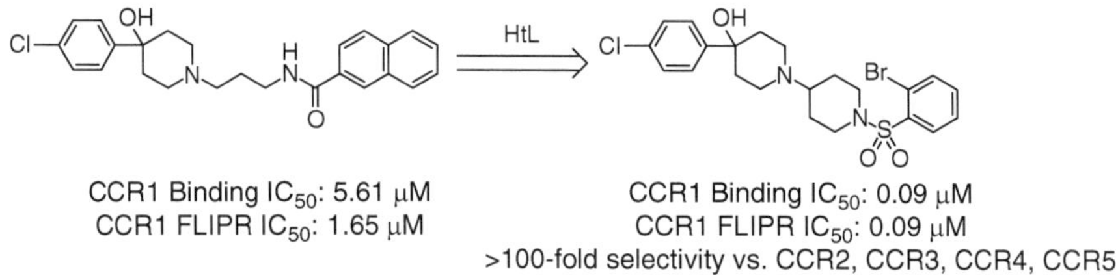

Scheme 23 CCR1 antagonists

to a 25 nM lead with excellent selectivity over a large panel of kinases (exceptions are IRK and TXR, also a member of the Tec family of kinases). While this novel series possesses excellent selectivity of a lead, the cellular potency of this series is weak and additional in vitro potency will need to be achieved to see better cellular activity (Scheme 22).

11.23 Example 23: CCR1 Antagonists

Chemokines play an important role in leukocyte migration and activation. Specifically, CCR1 has been implicated in the pathogenesis of chronic inflammatory diseases including rheumatoid arthritis and multiple sclerosis. Xie et al. [68] describe a series of 4-(4-chlorophenyl)piperidin-4-ols where the CCR1 potency/binding affinity was increased almost 100-fold by constraining the propyl amine moiety to a piperidine and changing the amide to a sulfonamide. The optimized compound has >100-fold selectivity vs CCR2, CCR3, CCR4, and CCR5 (Scheme 23).

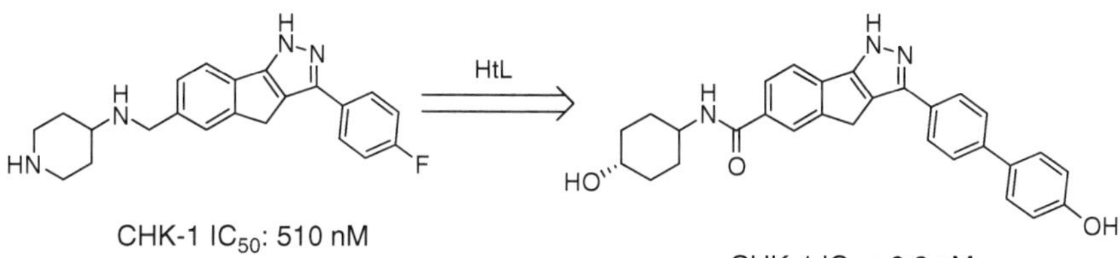

CHK-1 IC_{50}: 510 nM

CHK-1 IC_{50}: 6.2 nM
>200-fold selective for CHK-2, Aur1, CK2, PKA, PKCδ, PKCγ, EMK, ERK2, SGK, SRC, MARKAP2, AKT, CDC2
MTS assay with doxorubicin: EC_{50}: 1.8 μM
FACS assay with doxorubicin: EC_{50}: 0.77 μM
Camptothecin potentiation ratio: 6.7

Scheme 24 CHK-1 inhibitors

11.24 Example 24: CHK-1 Inhibitors

Recent studies have shown that inhibition of CHK-1 can make tumor cells sensitized to DNA-damaging chemotherapeutic agents, potentially making them more efficacious and selective. Tong et al. [69] have shown that a series of 1,4-dihydroindeno[1,2-*c*]pyrazole CHK-1 inhibitors can be optimized to a nM lead with >200-fold selectivity vs a number of related kinases. Moreover, the lead potentiates the effects of doxorubicin and camptothecin, both DNA-damaging agents in cell proliferation assays (MTS and soft agar assays) and abrogates G2/M checkpoint in a mechanism-based FACS assay. In vivo efficacy may be challenging since the hydroxyl groups should be easily conjugated and cleared via Phase II metabolic processes (Scheme 24).

12 Summary

High quality leads provide the foundation for the discovery of successful clinical development candidates, and therefore the identification of leads is an essential part of drug discovery. The thorough characterization of hits coming from a screening campaign is the important first step in lead identification. Once fully characterized, the hits with the highest likelihood of successful optimization to leads can be selected. Many factors contribute to the quality of a lead, including biological, physicochemical, ADME, and PK parameters. The identification of high quality leads, which are needed for successful lead optimization, requires the optimization of all of these parameters. Parallel optimization of all parameters is the most efficient way to achieve the goal of lead identification.

References

1. Alanine A, Nettekoven M, Roberts E, Thomas A (2003) Comb Chem High Throughput Screen 6:51
2. Davis AM, Keeling DJ, Steele J, Tomkinson NP, Tinker AC (2005) Curr Top Med Chem 5:421
3. Gillespie P, Goodnow RA Jr (2004) Annu Rep Med Chem 39:293
4. Goodnow RA Jr, Gillespie P (2007) Prog Med Chem 45:1
5. Keseru GM, Makara GM (2006) Drug Disc Today 11:741
6. Steinmeyer A (2006) ChemMedChem 1:31
7. Di L, Kerns E (2003) Curr Opin Chem Biol 7:402
8. Kerns E, Di L (2003) Drug Disc Today 8:316
9. Wang J, Faller B (2006) Compr Med Chem II 5:341
10. Feng BY, Simeonov A, Jadhav A, Babaoglu K, Inglese J, Shoichet BK, Austin CP (2007) J Med Chem 50:2385
11. Lundqvist T (2005) Curr Opin Drug Disc Dev 8:513
12. Lofas S (2004) Assay Drug Dev Technol 2:407
13. Jahnke W (2007) J Biomol NMR 39:87
14. Salvatella XG, Ernest G (2003) Chem Soc Rev 32:365–372
15. Scapin G (2006) Curr Pharm Des 12:2087
16. Nilakantan R, Nunn D, Greenblatt L, Walker G, Haraki K, Mobilio D (2006) J Chem Inf Model 46:1069
17. Schuffenhauer A, Brown N, Ertl P, Jenkins J, Selzer P, Hamon J (2007) J Chem Inf Model 47:325
18. Stahl M, Mauser H, Tsui M, Taylor N (2005) J Med Chem 48:4358
19. Davis A, Teague S (1999) Angew Chem Int Ed 38:736
20. Carhart R, Smith D, Venkataraghavan R (1985) J Chem Inf Comput Sci 25:64
21. Ellingboe J (2007) International Symposium on Advances in Synthetic and Medicinal Chemistry ASMC 07, St. Petersburg, Russia
22. Kansy MS, Gubernator K (1998) J Med Chem 41:1007
23. Alsenz J, Kansy M (2007) Adv Drug Deliv Rev 59:546
24. Di L, Kerns E, Hong Y, Chen H (2005) Int J Pharm 297:110
25. Wan H, Bergstroem F (2007) J Liq Chromatogr Relat Technol 30:681
26. Teague SD, Leeson PD, Oprea T (1999) Angew Chem Int Ed 38:3743
27. Lipinski CL, Dominy BW, Feeney PJ (1997) Adv Drug Deliv Rev 23:3
28. Veber DJ, Cheng HY, Smith BR, Ward KW, Kopple KD (2002) J Med Chem 45:2615
29. Morphy R (2006) J Med Chem 49:2969
30. Hopkins A, Groom C, Alex A (2004) Drug Disc Today 9:430
31. Reynolds CH, Tounge BA, Bembenek SD (2008) J Med Chem 51:2432
32. Chen Y, Monshouwer M, Fitch W (2007) Pharm Res 24:248
33. Nassar A-E, Kamel A, Clarimont C (2004) Drug Disc Today 9:1022
34. Jamieson C, Moir E, Rankovic Z, Wishart G (2006) J Med Chem 49:5029–5046
35. Aronov A (2006) J Med Chem 49:6917
36. Johnson S, Yue H, Conder M, Shi H, Doweyko A, Lloyd J, Levesque P (2007) Bioorg Med Chem 15:6182
37. Smith N, Raynaud F, Workman P (2007) Mol Cancer Ther 6:428
38. Korfmacher W, Cox K, Ng K, Veals J, Hsieh Y, Wainhaus S, Broske L, Prelusky D, Nomeir A, White R (2001) Rapid Commun Mass Spectrom 15:335
39. Di L, Kerns E, Fan K, McConnell O, Carter G (2003) Eur J Med Chem 38:223
40. Migliavacca E (2003) Mini Rev Med Chem 3:831
41. Eriksson L, Johansson E, Kettaneh-Wold N, Wold S (1999) Introduction to multi- and megavariate data analysis using projection methods (PCA and PLS). Umetrics AB, Umeå, Sweden

42. Benchware™ (2007), Data Miner, version 1.6; Tripos. St. Louis, Mo
43. Eriksson L, Johansson E, Kettaneh-Wold N, Wold S (2000) Design of experiments, principles and applications. Umetrics AB, Umeå, Sweden
44. Tye H (2004) Drug Disc Today 9:485
45. Gilbert AM, Kirisits M, Toy P, Nunn DS, Failli A, Dushin EG, Novikova E, Petersen PJ, Joseph-McCarthy D, McFadyen I, Fritz CC (2004) Bioorg Med Chem Lett 14:37
46. Zhao H (2007) Drug Discov Today 12:149
47. Wang X, Xu F, Xu Q, Mahmud H, Houze J, Zhu L, Akerman M, Tonn G, Tang L, McMaster BE, Dairaghi DJ, Schall TJ, Collins TL, Medina JC (2006) Bioorg Med Chem Lett 16:2800
48. McKew JC, Foley MA, Thakker P, Behnke ML, Lovering FE, Sum FW, Tam S, Wu K, Shen MW, Zhang W, Gonzalez M, Liu S, Mahadevan A, Sard H, Khor SP, Clark JD (2006) J Med Chem 49:135
49. Souers AJ, Wodka D, Gao J, Lewis JC, Vasudevan A, Gentles R, Brodjian S, Dayton B, Ogiela CA, Fry D, Hernandez LE, Marsh KC, Collins CA, Kym PR (2004) Bioorg Med Chem Lett 14:4873
50. Lund BW, Piu F, Gauthier NK, Eeg A, Currier E, Sherbukhin V, Brann MR, Hacksell U, Olsson R (2005) J Med Chem 48:7517
51. Beaulieu PL, Bos M, Bousquet Y, Fazal G, Gauthier J, Gillard J, Goulet S, LaPlante S, Poupart MA, Lefebvre S, McKercher G, Pellerin C, Austel V, Kukolj G (2004) Bioorg Med Chem Lett 14:119
52. Williams TM, Stump CA, Nguyen DN, Quigley AG, Bell IM, Gallicchio SN, Zartman CB, Wan B-L, Della Penna K, Kunapuli P, Kane SA, Koblan KS, Mosser SD, Rutledge RZ, Salvatore C, Fay JF, Vacca JP, Graham SL (2006) Bioorg Med Chem Lett 16:2595
53. Morwick T, Berry A, Brickwood J, Cardozo M, Catron K, DeTuri M, Emeigh J, Homon C, Hrapchak M, Jacober S, Jakes S, Kaplita P, Keıly TA, Ksiazek J, Liuzzi M, Magolda R, Mao C, Marshall D, McNeil D, Prokopowicz A, III, Sarko C, Scouten E, Sledziona C, Sun S, Watrous J, Wu JP, Cywin CL (2006) J Med Chem 49:2898
54. Baxter A, Brough S, Cooper A, Floettmann E, Foster S, Harding C, Kettle J, McInally T, Martin C, Mobbs M, Needham M, Newham P, Paine S, St-Gallay S, Salter S, Unitt J, Xue Y (2004) Bioorg Med Chem Lett 14:2817
55. Cywin CL, Dahmann G, Prokopowicz AS, IIIYoung ER, Magolda RL, Cardozo MG, Cogan DA, Disalvo D, Ginn JD, Kashem MA, Wolak JP, Homon CA, Farrell TM, Grbic H, Hu H, Kaplita PV, Liu LH, Spero DM, Jeanfavre DD, O'Shea KM, White DM, Woska JR Jr, Brown ML (2007) Bioorg Med Chem Lett 17:225
56. Poulain R, Horvath D, Bonnet B, Eckhoff C, Chapelain B, Bodinier MC, Deprez B (2001) J Med Chem 44:3391
57. Palmer MJ, Bell AS, Fox DN, Brown DG (2007) Curr Top Med Chem 7:405
58. Baxter A, Bennion C, Bent J, Boden K, Brough S, Cooper A, Kinchin E, Kindon N, McInally T, Mortimore M, Roberts B, Unitt J (2003) Bioorg Med Chem Lett 13:2625
59. Baxter A, Cooper A, Kinchin E, Moakes K, Unitt J, Wallace A (2006) Bioorg Med Chem Lett 16:960
60. Ho K-K, Auld DS, Bohnstedt AC, Conti P, Dokter W, Erickson S, Feng D, Inglese J, Kingsbury C, Kultgen SG, Liu R-Q, Masterson CM, Ohlmeyer M, Rong Y, Rooseboom M, Roughton A, Samama P, Smit M-J, Son E, Van der Louw J, Vogel G, Webb M, Wijkmans J, You M (2006) Bioorg Med Chem Lett 16:2724
61. Armour D, de Groot MJ, Edwards M, Perros M, Price DA, Stammen BL, Wood A (2006) ChemMedChem 1:706
62. Pevarello P, Brasca MG, Amici R, Orsini P, Traquandi G, Corti L, Piutti C, Sansonna P, Villa M, Pierce BS, Pulici M, Giordano P, Martina K, Fritzen EL, Nugent RA, Casale E, Cameron A, Ciomei M, Roletto F, Isacchi A, Fogliatto G, Pesenti E, Pastori W, Marsiglio A, Leach KL, Clare PM, Fiorentini F, Varasi M, Vulpetti A, Warpehoski MA (2004) J Med Chem 47:3367
63. Baxter A, Bent J, Bowers K, Braddock M, Brough S, Fagura M, Lawson M, McInally T, Mortimore M, Robertson M, Weaver R, Webborn P (2003) Bioorg Med Chem Lett 13:4047

64. Backes BJ, Longenecker K, Hamilton GL, Stewart K, Lai C, Kopecka H, von Geldern TW, Madar DJ, Pei Z, Lubben TH, Zinker BA, Tian Z, Ballaron SJ, Stashko MA, Mika AK, Beno DW, Kempf-Grote AJ, Black-Schaefer C, Sham HL, Trevillyan JM (2007) Bioorg Med Chem Lett 17:2005
65. Cole DC, Stock JR, Chopra R, Cowling R, Ellingboe JW, Fan KY, Harrison BL, Hu Y, Jacobsen S, Jennings LD, Jin G, Lohse PA, Malamas MS, Manas ES, Moore WJ, O'Donnell M-M, Olland AM, Robichaud AJ, Svenson K, Wu J, Wagner E, Bard J (2008) Bioorg Med Chem Lett 18:1063
66. Wang X, Kolasa T, El Kouhen OF, Chovan LE, Black-Shaefer CL, Wagenaar FL, Garton JA, Moreland RB, Honore P, Lau YY, Dandliker PJ, Brioni JD, Stewart AO (2007) Bioorg Med Chem Lett 17:4303
67. Snow RJ, Abeywardane A, Campbell S, Lord J, Kashem MA, Khine HH, King J, Kowalski JA, Pullen SS, Roma T, Roth GP, Sarko CR, Wilson NS, Winters MP, Wolak JP, Cywin CL (2007) Bioorg Med Chem Lett 17:3660
68. Xie YF, Sircar I, Lake K, Komandla M, Ligsay K, Li J, Xu K, Parise J, Schneider L, Huang D, Liu J, Sakurai N, Barbosa M, Jack R (2008) Bioorg Med Chem Lett 18:2215
69. Tong Y, Claiborne A, Stewart KD, Park C, Kovar P, Chen Z, Credo RB, Gu WZ, Gwaltney IS, Judge RA, Zhang H, Rosenberg SH, Sham HL, Sowin TJ, Lin Nh (2007) Bioorg Med Chem 15:2759
70. Poulain R, Horvath D, Bonnet B, Eckhoff C, Chapelain B, Bodinier MC, Deprez B (2001) J Med Chem 44:3378

Index